Engineering Capstone Design

Engineering Capstone Design

Bahram Nassersharif

CRC Press
Taylor & Francis Group
Boca Raton London New York

CRC Press is an imprint of the
Taylor & Francis Group, an **informa** business

First edition published 2022
by CRC Press
6000 Broken Sound Parkway NW, Suite 300, Boca Raton, FL 33487-2742

and by CRC Press
2 Park Square, Milton Park, Abingdon, Oxon, OX14 4RN

© 2022 Taylor & Francis Group, LLC

CRC Press is an imprint of Taylor & Francis Group, LLC

Reasonable efforts have been made to publish reliable data and information, but the author and publisher cannot assume responsibility for the validity of all materials or the consequences of their use. The authors and publishers have attempted to trace the copyright holders of all material reproduced in this publication and apologize to copyright holders if permission to publish in this form has not been obtained. If any copyright material has not been acknowledged please write and let us know so we may rectify in any future reprint.

Except as permitted under U.S. Copyright Law, no part of this book may be reprinted, reproduced, transmitted, or utilized in any form by any electronic, mechanical, or other means, now known or hereafter invented, including photocopying, microfilming, and recording, or in any information storage or retrieval system, without written permission from the publishers.

For permission to photocopy or use material electronically from this work, access www.copyright.com or contact the Copyright Clearance Center, Inc. (CCC), 222 Rosewood Drive, Danvers, MA 01923, 978-750-8400. For works that are not available on CCC please contact mpkbookspermissions@tandf.co.uk

Trademark notice: Product or corporate names may be trademarks or registered trademarks and are used only for identification and explanation without intent to infringe.

ISBN: 978-0-367-62159-9 (hbk)
ISBN: 978-0-367-62161-2 (pbk)
ISBN: 978-1-003-10821-4 (ebk)

DOI: 10.1201/9781003108214

Typeset in Times
by codeMantra

Access the Support Material: www.routledge.com/9780367621599

Contents

Preface xi
Author xiii

PART I Define, Conceive, Prove, Document

Chapter 1   Engineering Design 3

  1.1   Introduction 3
  1.2   Engineering Design Process 5
  1.3   Design and Engineering Accreditation 6
  1.4   Open-Ended Design Problems 10
  1.5   Regulations, Codes, and Standards 11
        1.5.1   Regulations 11
        1.5.2   Codes and Standards 13
        1.5.3   Specifications 16
  1.6   Capstone Design Process 16
  1.7   Preparing for Capstone Design 21

Chapter 2   Design Project Team 23

  2.1   Selection of Team Members for Design Projects 24
        2.1.1   Background Research on the Project 25
        2.1.2   Educational Preparation 25
        2.1.3   Professional Preparation 26
        2.1.4   Qualifications 26
        2.1.5   Professionalism in Interactions 27
        2.1.6   Team Dynamics 28
                2.1.6.1   Effective Team Membership 29
                2.1.6.2   Team Leadership 30
        2.1.7   Roles and Responsibilities 30
        2.1.8   Effective Team Management 30
                2.1.8.1   Scheduling 31
                2.1.8.2   Agenda and Time Management 31
                2.1.8.3   Minutes 32
                2.1.8.4   Action Items 32
        2.1.9   Progress Reports 32
        2.1.10 Communications 33
        2.1.11 Online Communication 35
                2.1.11.1   Online File Management 36
                2.1.11.2   Real-Time Conferencing Tools 36

v

Chapter 3 Project Management 39

- 3.1 Planning 41
- 3.2 Scheduling 41
- 3.3 Entering Project Information into MS Project 43
  - 3.3.1 Independent and Dependent Tasks 44
- 3.4 Assigning and Accepting Responsibility 45
- 3.5 Working with the Project Plan 49
  - 3.5.1 Resources 49
  - 3.5.2 Gantt Chart 50
  - 3.5.3 Network Diagram and Optimizing the Project Plan 51
- 3.6 Resource Leveling 56
- 3.7 Project Reports 57

Chapter 4 Defining the Design Problem 63

- 4.1 Methods for Understanding Open-Ended Problems 63
  - 4.1.1 Brainstorming 64
- 4.2 Gathering Information 65
  - 4.2.1 Surveys 66
- 4.3 Sources of Information 68
  - 4.3.1 Literature Search 70
    - 4.3.1.1 Literature Search Assignment 71
  - 4.3.2 Patents and Patent Search 72
    - 4.3.2.1 Patent Search Assignment 75
- 4.4 Design for X 76
- 4.5 Design Specifications 78
  - 4.5.1 Customer Needs and Requirements 80
  - 4.5.2 Design Specifications Leading Questions 80
  - 4.5.3 Design Specifications Assignment 81

Chapter 5 Conceiving Design Solutions 83

- 5.1 Creative Thinking 83
  - 5.1.1 Fostering Creative Thinking 84
  - 5.1.2 Barriers to Creative Thinking 84
  - 5.1.3 Techniques for Creative Thinking and Brainstorming 85
    - 5.1.3.1 Technique 1: Silent Brainstorming 86
    - 5.1.3.2 Technique 2: Evolutionary Thinking – The TRIZ Method 86
    - 5.1.3.3 Technique 3: Random Input 86
    - 5.1.3.4 Technique 4: SCAMPER 86
- 5.2 Generating Solution Concepts 88
  - 5.2.1 Functional Decomposition 88
  - 5.2.2 Morphological Methods 90

Contents   vii

         5.2.3   Axiomatic Design 90
         5.2.4   TRIZ Method 90
                 5.2.4.1   Contradictions 91
   5.3   Analyzing Concepts 94
         5.3.1   Engineering Analysis 94
         5.3.2   Modeling 95
         5.3.3   Simulation 95
   5.4   Decision-Making 95
         5.4.1   Pugh Chart 96
         5.4.2   Quality Function Deployment 97
         5.4.3   QFD Analysis Example 100

Chapter 6   Critical Design Review 107
   6.1   Engineering Team Project Presentation 107
   6.2   Receiving Critique 107
   6.3   Incorporating Feedback 108

Chapter 7   Proof of Concept 109
   7.1   How to Create a Proof of Concept 110
         7.1.1   Demonstrate the Design Solution Meets
                   Sponsor Requirements 110
         7.1.2   Generate an Improved Design Solution 110
         7.1.3   Build a Prototype and Test It 111
         7.1.4   Collect Test Data, Analyze, and Document 111
         7.1.5   POC Presentation 111
         7.1.6   Viability and Usability 112

Chapter 8   Documentation 113
   8.1   Engineering Design Documentation 113
         8.1.1   Engineering Logbook 114
         8.1.2   Design Binder 116
         8.1.3   Electronic Files and Project Archive 118
   8.2   Verbal Presentation with Slides 119
         8.2.1   Guidelines for Capstone Design Presentations 119
   8.3   Photos 129
   8.4   Video Presentation 130
         8.4.1   Timing Studies 130
   8.5   Poster 131
         8.5.1   Guidelines for Posters 131
                 8.5.1.1   How Much Poster Space Are You
                            Allowed? 131
                 8.5.1.2   Format 132
                 8.5.1.3   Planning 132
                 8.5.1.4   Content 133
                 8.5.1.5   Design 133

	8.6	Technical Information Sheet	136
		8.6.1 Guidelines for Capstone Technical Information Sheets	136
		8.6.2 Style Guide for Capstone Technical Information Sheet	137
	8.7	Preliminary Design Report	137
		8.7.1 Guidelines for Preliminary Design Report	137
	8.8	Final Design Report	142
		8.8.1 Guidelines for the Final Design Report	142

PART II Build, Test, Redesign, Repeat, Document

Chapter 9 Building from Proof of Concept Design 151

	9.1	Logistics of Building the Design	151
	9.2	Resources Needed	152
	9.3	Actual Versus Model	154
	9.4	Materials	154
		9.4.1 Materials Selection Charts	155
		9.4.2 Metals	158
		9.4.3 Plastics	160
	9.5	Manufacturability	162
		9.5.1 Manufacturing Process Selection	163
		9.5.2 Design for Manufacturing and Assembly (DFMA)	163
		9.5.3 Mistake Proofing	165
	9.6	3D Printing	167
	9.7	Quality	172
	9.8	Bill of Materials	173
	9.9	Procurement	173

Chapter 10 Redesign 175

	10.1	Application of Test Results	176
	10.2	Adjustments to Design-Build	176
	10.3	Major Design Changes	177
	10.4	Optimization	177
	10.5	Quality Engineering	178
	10.6	Cost Evaluation	179
	10.7	Return on Investment	179
	10.8	Design Optimization	180

Contents

Chapter 11 Closing Out the Project and Documentation 183
      11.1   Planning a Final Capstone Event 183
            11.1.1 Estimating the Audience Size and Selecting a Venue 183
            11.1.2 Select a Date 184
            11.1.3 Select a Format 184
            11.1.4 Create a Program 184
            11.1.5 Promote the Event and Invite People 185
            11.1.6 Invite Media 185
      11.2   Design Presentation to Sponsor 186

Bibliography 187

Index 193

Preface

University engineering programs in the United States have experienced major transitions in the curriculum and student outcomes requirements for the past three decades. Since ABET 2000, engineering programs have been required to demonstrate a major design experience for their students. The major design experience is commonly called Senior Engineering Capstone Design, Senior Engineering Design, or Capstone Design. Capstone Design is the culmination of the engineering curriculum. It is a required element of the accredited engineering curricula where students must engage in a design experience utilizing their knowledge and skills acquired in earlier years and course work. ABET accreditation is also growing internationally as a rigorous standard for quality of engineering programs around the world. ABET defines Engineering Design as:

> Engineering design is a process of devising a system, component, or process to meet desired needs and specifications within constraints. It is an iterative, creative decision-making process in which the basic sciences, mathematics, and engineering sciences are applied to convert resources into solutions. Engineering design involves identifying opportunities, developing requirements, performing analysis and synthesis, generating multiple solutions, evaluating solutions against requirements, considering risks, and making tradeoffs for the purpose of obtaining a high-quality solution under the given circumstances. For illustrative purposes only, examples of possible constraints include accessibility, aesthetics, codes, constructability, cost, ergonomics, extensibility, functionality, interoperability, legal considerations, maintainability, manufacturability, marketability, policy, regulations, schedule, standards, sustainability, or usability.

This book is intended for instructors and students in Capstone Design courses to guide their engagement in the design projects in a step-wise and logical manner regardless of their engineering discipline. The book is structured to accommodate a two-semester Capstone Design sequence, incorporating the accreditation requirements and providing a modern framework for working with the industry. The layout of the topics has been developed and continuously improved over a decade of teaching capstone design. This book should also be of interest to professionals working in the industry engaged in product development, design problem-solving, and sponsoring capstone design projects.

Author

Bahram Nassersharif is a Distinguished University Professor in the Department of Mechanical, Industrial, and Systems Engineering at the University of Rhode Island. He received his B.S. in Mathematics and Ph.D. in Nuclear Engineering from Oregon State University. He was the youngest graduate of Oregon State University in 1980 among the class of 3,300 graduates. He was the youngest person in the U.S.A. to complete his Ph.D. in engineering in 1982. After a period at Los Alamos National Laboratory, he joined Texas A&M University (TAMU) as an Assistant Professor and founding Director of the TAMU Supercomputer Center. He moved to the University of Nevada in 1991 as the Director of the National Supercomputing Center for Energy and the Environment and Professor of Mechanical Engineering. In 1997, he joined New Mexico State University as a Professor and Department Head of Mechanical Engineering. He became Dean of Engineering at the University of Rhode Island in 2003. Since 2007, he has developed, continually improved, and taught the senior Capstone Design sequence in Mechanical Engineering. Professor Nassersharif is a recipient of the Presidential Young Investigator Award (1986) and a fellow of the AAAS. He is a member of ASME, ANS, IEEE, IEEE-CS, ACM, and AAAI. His current research and teaching interests are in engineering design, modeling and simulation, nuclear systems, and space nuclear power and propulsion.

Part I

*Define, Conceive,
Prove, Document*

1 Engineering Design

1.1 INTRODUCTION

This book is for students, mentors, sponsors, and professors engaged in engineering capstone design. It is not meant for a specific discipline of engineering but rather a comprehensive approach to capstone design.

Many different disciplines use the word design to describe some methods of problem-solving in that particular area. If you look at the dictionary definition of design, as a verb, we see definitions such as to devise, contrive, to have as a purpose, to devise for a specific function or end, to make a drawing, pattern, or sketch of, and to draw the plans for. Design as a noun has dictionary listings such as a mental project or scheme in which means to an end are laid down, a preliminary sketch or outline showing the main features of something to be executed, the arrangement of elements or details in a product or work of art, and the creative art of executing aesthetic or functional designs. None of these definitions fully captures the practice and the meaning of design in engineering.

Engineering design, sometimes known as the engineering method, is a formal, rigorous, and systematic process to optimize a problem. Problems are often expressed as a desire to solve a situation that has not been solved before or improvements on something that already exists, whether it's a process, a device, or a concept. Consequently, by the very nature of this statement of problems, they are incomplete and ill-defined. The engineering design process starts with defining and understanding the problem and what is to be achieved.

Figure 1.1 shows the process diagram for the engineering method (also known as engineering design).

Once a problem has been presented, then it is necessary to do better understand the problem. Additional information about the problem area and related topics can be gathered through an online search of the literature, related patents, and competitor information. The problem-solving engineer is responsible for developing the design specifications and any constraints that might be appropriate. Solutions are then generated in the form of concepts that satisfy the design specifications and the constraints. The concept may be a device or a process. If the solution is a device, then it may be possible to prototype it. If the concept is a process, it can be modeled perhaps in a software model. The prototype or the model can be tested against the design specifications, constraints, and any additional considerations such as ethical and environmental. The test results are then used to assess the fitness of the solution concept to the problem. Iterations on the concept, prototype, and model are used to improve, optimize, or streamline the solution. Once a satisfactory solution is achieved, it is documented, communicated, and delivered to the sponsor of the problem or project. The engineering design method is unique to engineering, and it distinguishes the work of engineers from all other forms of professionals, including

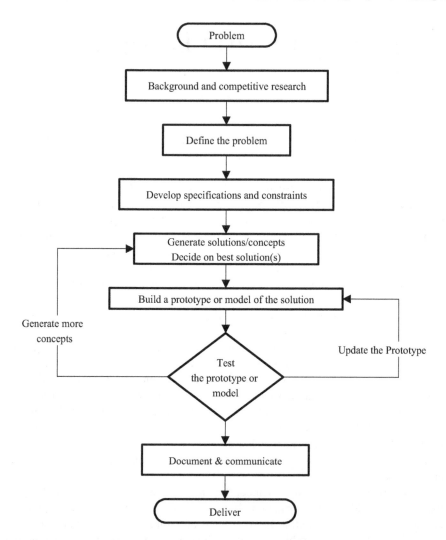

FIGURE 1.1 The engineering method.

scientists and mathematicians. We demonstrate this difference by comparing the engineering method to the scientific method.

Science is an essential constituent of engineering. Sometimes engineering is confused with applied science. Sciences rely on the scientific method as their fundamental tool for discovery and advancement. Scientists develop hypotheses based on observations of natural phenomena. A hypothesis is posed to describe or encapsulate the observation based on fundamental questions (formulated by the scientist) about their observations. Figure 1.2 shows the flowchart for the scientific method.

Scientists formulate experiments to test their hypotheses. The experiments may answer their questions directly and confirm their hypothesis, or it may reveal additional considerations or issues that may emerge from the results. They also attempt to

Engineering Design

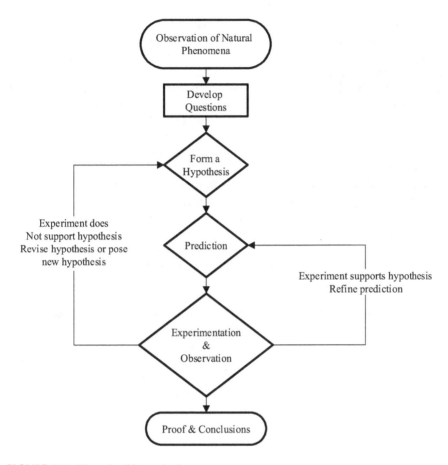

FIGURE 1.2 The scientific method.

apply their theory or hypothesis to other situations and observations similar to their original observations. Often, experiments result in a need to modify the hypothesis or constrain it to be more directly applicable. If the iteration results on hypothesis, prediction, and testing converge after a sufficient number of iterations, the theory or hypothesis may be proven to be true, false, or somewhat true.

Scientific discovery relies on peer review. Suppose peers can repeat the same experiments and design their experiments to validate the hypothesis further. In that case, the proven hypothesis becomes more widely accepted, and eventually, engineers use it to design systems and devices based on those scientific discoveries and proven principles.

1.2 ENGINEERING DESIGN PROCESS

The engineering method described in the previous section is the foundation of an engineering design process. Additional considerations come into play for an engineering design activity, such as cost and schedules as explicit components that must be included in an industry or professional setting. Often cost and delivery times become the major decision-making factors for the industry design projects.

Some concepts will take longer, or they may cost more, resulting in shelving or eliminating those ideas. It is imperative in the engineering design process to estimate costs and timelines as accurately as possible. The process often includes a proposal, resulting in a legal contract between the design engineers and the customers or sponsors. The contractual requirements, pricing, and schedules may constrain the design engineers to take more conservative approaches to be sure to deliver on time and within budget. The design problem and approach to solving it become heavily influenced by the cost and schedule factors.

We can update our flowchart to include this important element of the real-world engineering design into our problem-solving process.

Figure 1.3 shows the engineering design process flowchart. In this process, all efforts and costs associated with developing the design project proposal are burdened by the engineering design group. Those costs are then recovered in the form of an overhead or indirect cost on the project for each proposal that does receive funding support.

The cost and schedule considerations are part of the responsibility of every engineer on the team, their methodology for engineering problem-solving based on their experience, knowledge, and skills. Engineers who are successful in preparing winning project proposals and successfully completing the work they proposed will thrive. Those who fail in this process will work under the direction of others. The project proposal preparation, accurate costing, and meeting project timeline are prerequisites to having the opportunity to do excellent technical work on the design. These same elements apply to engineering capstone design, which we will cover later in this chapter.

Because of the central importance of design in engineering practice, it has been required by Accreditation Board for Engineering and Technology (ABET) as part of the curricula in engineering.

1.3 DESIGN AND ENGINEERING ACCREDITATION

Engineering accreditation in the United States started with the Engineer's Council for Professional Development (EPCD) as a professional organization "dedicated to the education, accreditation, regulation, and professional development of engineering professionals and students." EPCD was renamed the Accreditation Board for Engineering and Technology in 1980. Starting in 2005, ABET was incorporated as Accreditation Board for Engineering and Technology, Inc. and used only the acronym ABET. This change to just ABET was in part in response to the significant increase in international interest, outside of the United States, to adopt ABET philosophy and accreditation requirements in engineering programs [https://www.abet.org/about-abet/history/].

After many years of work, deliberations, and engineering community input, ABET adopted a revolutionary set of changes to its accreditation requirements and process in 1997, named Criteria 2000 (EC2000). The big change was in the focus of accreditation on what the engineering students learned rather than the previous focus

Engineering Design

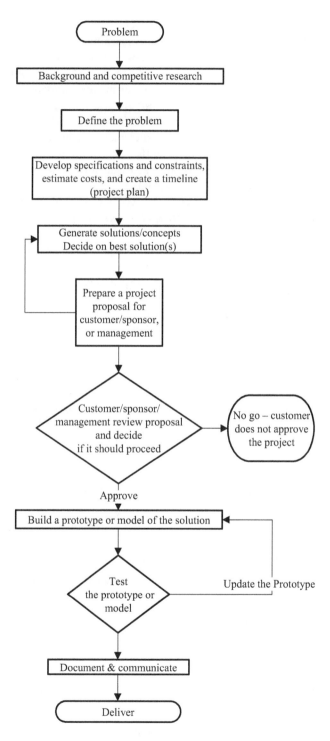

FIGURE 1.3 The engineering design process.

on what was taught in the program. EC2000 required that engineering programs establish published education objectives, student learning outcomes, and assessment processes to ensure that the engineering program curriculum provides the technical and professional skills that employers need and demand. Also, the engineering program must demonstrate that the faculty are measuring and assessing student learning outcomes. EC2000 empowered engineering programs to take responsibility for their curriculum and engineering education of their students and demonstrate that to ABET. Later, ABET introduced the additional element of continuous improvement. The engineering programs also had to show that they are improving their program and processes for student learning outcomes assessment by collecting data, analyzing, and taking appropriate improvement actions.

So, who is ABET? The short answer is that ABET is us. ABET comprises representatives of the engineering (also applied sciences and computer science) profession through recognized and established national engineering professional organizations. Today, ABET accredits 4,144 programs at 812 colleges and universities in 32 countries. The engineering accreditation committee (EAC) is responsible for creating criteria for accrediting engineering programs and organizing the review of engineering programs by training and deploying review teams for the evaluation of programs. The criteria are revised based on lessons learned during the review process, and changes are recommended by the professional organizations and ABET committees to adjust or edit the criteria. Table 1.1 shows the engineering accreditation criteria for 2020–2021.

The student outcomes criterion defines what students should know but more specifically defines student outcomes for engineering design. Engineering students should become familiar with these criteria because they are part of the overall evaluation system for engineering programs. Particularly, students should know the educational objectives of their program of study and what is expected of them concerning learning outcomes. The educational objectives for the engineering program should be publicly available on the program website. Table 1.2 shows the student outcomes outlined under Criterion 3 of the engineering accreditation criteria for 2020–2021.

TABLE 1.1
Engineering Accreditation Criteria 2020–2021

The Engineering Accreditation Criteria 2020–2021 for undergraduate engineering programs consist of nine categories as follows:
- Criterion 1: Students
- Criterion 2: Program Educational Objectives
- Criterion 3: Student Outcomes
- Criterion 4: Continuous Improvement
- Criterion 5: Curriculum
- Criterion 6: Faculty
- Criterion 7: Facilities
- Criterion 8: Institutional Support
- Program Specific Criteria (established by lead society)

TABLE 1.2
ABET 2020–2021 Engineering Accreditation Criterion 3

Criterion 3. Student Outcomes: The program must have documented student outcomes that support the program's educational objectives. Attainment of these outcomes prepares graduates to enter the professional practice of engineering. Student outcomes are outcomes (1) through (7), plus any additional outcomes that the program may articulate.

Outcome 1:	an ability to identify, formulate, and solve complex engineering problems by applying principles of engineering, science, and mathematics
Outcome 2:	an ability to apply engineering design to produce solutions that meet specified needs with consideration of public health, safety, and welfare, as well as global, cultural, social, environmental, and economic factors
Outcome 3:	an ability to communicate effectively with a range of audiences
Outcome 4:	an ability to recognize ethical and professional responsibilities in engineering situations and make informed judgments, which must consider the impact of engineering solutions in global, economic, environmental, and societal contexts
Outcome 5:	an ability to function effectively on a team whose members together provide leadership, create a collaborative and inclusive environment, establish goals, plan tasks, and meet objectives
Outcome 6:	an ability to develop and conduct appropriate experimentation, analyze and interpret data, and use engineering judgment to draw conclusions
Outcome 7:	an ability to acquire and apply new knowledge as needed, using appropriate learning strategies

These criteria are periodically reviewed and revised through an engineering community-based process.

All seven outcomes apply to the capstone design experience. Many engineering programs use capstone design courses as an essential assessment tool in evaluating the effectiveness of their engineering curricula. The capstone design experience requirement is under Outcome 5(d), which states:

The curriculum must include a culminating major engineering design experience that (a) incorporates appropriate engineering standards and multiple constraints and (b) is based on the knowledge and skills acquired in earlier course work.

Engineering programs must show that they are achieving the above requirements for the student outcomes and curriculum. They do so by preparing a self-study document for the ABET review. The review cycles are 6 years for fully accredited programs without any weaknesses or deficiencies noted during the previous review cycle.

If a program has never been accredited and preparing for an accreditation review for the first time, they will need to prepare and have the following in place before starting the process. The program must have a sufficient number of graduates to show that it has established a successful track record in educating engineering students. The program will show that graduates have either been employed in an appropriate industry or government sector or have pursued graduate studies. The new program must have engaged in assessing their curriculum and their student learning outcomes as outlined by ABET requirements. It will be helpful to the program to demonstrate a continuous improvement process. The program must also have a sufficient number of qualified faculty and provide institutional resources to conduct its mission successfully.

Programs that have been accredited previously start their work on the next cycle of ABET review shortly after receiving their review report from ABET. The process of the review by ABET starts with a notification to ABET by the dean of engineering that specific engineering programs are coming up for review approximately 2 years before the end of the accreditation period for those programs.

The review process begins with the program preparing for the accreditation requirements either as a completely new review for a program (that has never been reviewed before) or a review of a previously accredited program on a 6-year cycle. The results of the accreditation reviews are communicated in writing to the leadership of the institution and the program.

1.4 OPEN-ENDED DESIGN PROBLEMS

Engineering programs do an excellent job of educating engineering students on becoming excellent problem-solvers. Traditional textbooks in engineering reinforce the materials covered in each chapter of the book by having exercise problems. Those problems are designed in such a way that the students must apply the knowledge they gained in reading the chapter and apply it to those problems. The exercise problems have one correct answer. Consequently, engineering students are trained to think of problems having a unique solution with a limited set of options for solving the problem. While this is a rigorous method to assure student understanding of the chapter of the materials they read, it does not directly relate to the real world. Engineering students struggle with the vast space that surrounds real-world engineering problems. We call these types of design problems open-ended. For open-ended problems, solutions are much harder to constrain and construct.

Open-ended problems are more satisfying and fun to work on, particularly as a team. You have to participate actively in problem-solving and design and express your ideas more frequently. The description provided for an open-ended problem by its very nature is incomplete and ill-posed. You will have to develop the problem's definition by asking many questions related to what is provided in the problem statement.

Typically, open-ended problems have not been solved before. Consequently, the person posing the problem also does not know the solution to the problem, nor can they provide additional details needed to create a complete problem definition. Open-ended problems require an investigative approach to gathering additional information required to start the problem-solving process. Not having complete information about the problem creates a paradox for the engineering student because, until now, we had focused on problem-solving in the engineering curriculum. You cannot move forward until you have taken the steps necessary to define the problem. You cannot just ask the sponsor of the problem because, in most cases, the sponsor does not have complete information. They depend on you as the design engineer to find the information that you need to create a design solution that will satisfy the sponsor.

Engineering Design 11

So, what information do you need? Where do you find that additional information? Could this problem have been solved before? Was there something similar designed previously? Are there solutions that are provided commercially, as a product or service? Who are the experts in that field of study? Do you have access to those experts? What type of resources do you need to find additional information? Is the problem too complex, or do you not feel you have the technical knowledge to attempt it? Will it be possible, practical, or feasible to create a solution to the problem? What do you know about the problem? What don't you know about the problem? Are there any non-technical considerations in achieving a design solution? Who might be interested in your design solution? Will the problem and your design solution meet the requirements that your professor has established for the capstone design course? Are you looking for the best design solution to the problem or any design solution?

You will have to answer the above questions and many more that you will develop on your own related to your project. You will have to document the process that you used to find the information and cite the sources. We will discuss the steps that will help you achieve a sufficiently complete definition of your design problem in the chapters that follow.

1.5 REGULATIONS, CODES, AND STANDARDS

All design projects that result in people using the product or process created by the designers will be subject to some regulations, codes, standards, or specifications. The purpose of the regulations, codes, standards, and specifications is to protect people from harm, create norms so that products and services are compatible, and establish safety and quality measures.

1.5.1 REGULATIONS

Governments create laws and regulations to protect the health and welfare of their citizens, the structure of their societies, and defend their rights and territories. All governments have regulatory bodies and agencies and further define rules of practice consistent with the laws established in that country for governing the creation, sales, service, delivery, and utilization of devices, processes, and services by their citizens. For example, in the United States, the following agencies were created by the Federal Government to regulate (and enforce laws) appropriate industries or services:

- **Consumer Product Safety Commission (CPSC)** – enforces federal safety standards
- **Environmental Protection Agency (EPA)** – establishes and enforces pollution standards
- **Equal Employment Opportunity Commission (EEOC)** – administers and enforces Title VIII or the Civil Rights Act of 1964 (fair employment)
- **Federal Aviation Administration (FAA)** – regulates and promotes air transportation safety, including airports and pilot licensing
- **Federal Communications Commission (FCC)** – regulates interstate and foreign communication by radio, telephone, telegraph, and television

- **Federal Deposit Insurance Corporation (FDIC)** – insures bank deposits, approves mergers, and audits banking practices
- **Federal Reserve System (the FED)** – regulates banking and manages the money supply
- **Federal Trade Commission (FTC)** – ensures free and fair competition and protects consumers from unfair or deceptive practices
- **Food and Drug Administration (FDA)** – administers federal food purity laws, drug testing and safety, and cosmetics
- **Interstate Commerce Commission (ICC)** – enforces federal laws concerning transportation that crosses state lines
- **National Institute of Standards and Technology (NIST)** – establish and maintain standards in measurement and technological developments
- **National Labor Relations Board (NLRB)** – prevents or corrects unfair labor practices by either employers or unions
- **Nuclear Regulatory Commission (NRC)** – licenses and regulates non-military nuclear facilities
- **Occupational Safety and Health Administration (OSHA)** – develops and enforces federal standards and regulations ensuring working conditions
- **Securities and Exchange Commission (SEC)** – administers federal laws concerning the buying and selling of securities

The above agencies provide certification or licensing processes for enforcing laws and regulating the conduct of individuals and companies to assure compliance with laws and regulations. They publish their regulations, usually in the form of the Code of Federal Regulations, and solicit public comment and feedback in their regulatory processes. Your design project may be directly subject to some of the regulations enforced by the Federal Regulatory agencies, or your design project may be part of a larger system or process sponsored by a company that may be subject to those regulations. You can find the appropriate regulations easily by searching the internet for the appropriate areas. For example, many design projects may be subject to multiple regulations overseen by CPSC, EPA, and OSHA as a minimum. Projects related to aeronautics and astronautics may be subject to FAA regulations. Projects in nuclear engineering may be subject to NRC regulations. Projects related to food and medical applications will be subject to FDA regulations.

State governments also have regulatory bodies or authorities supervising the activities of their citizens, residents, visitors, and businesses in their territories. They are created by the state governments to regulate and enforce laws for safety, compliance with standards and codes, and protecting consumers. Regulatory bodies are typically part of the executive branch of the government. They focus on areas that are complex and require supervision without entanglement in the normal political process of lawmaking. They deal with the following areas:

- Advertising regulation
- Alcoholic beverages
- Bank regulation
- Cable and internet services

Engineering Design

- Children's safety and well-being
- Communication services
- Consumer protection
- Cyber-security regulation
- Economic regulation
- Electricity generation, distribution, and pricing
- Environmental regulation
- Financial regulation
- Food safety and food security
- Noise regulation
- Radiation safety
- Minerals
- Occupational safety and health
- Public health
- Regulation and monitoring of pollution
- Regulation of acupuncture
- Regulation of nanotechnology
- Regulation of sport
- Regulation of therapeutic goods
- Regulation through litigation
- Vehicle regulation
- Regulation of ship pollution in the United States
- Regulation and prevalence of homeopathy
- Regulation of science
- Roads and transportation
- Wage regulation

Your design projects may be regulated by a state agency in addition to the federal agency. The state regulatory authorities publish their requirements and enforce their areas of responsibility through inspection, registration, certification, and licensing. You can look up this information online.

You can look up similar regulatory information for other countries and regions of the world as appropriate for your design project. For example, if you are designing a product that will be sold in many countries, that product will be subject to regulations in all of those countries. Different versions of products or services are frequently created that are compliant with regulations in the countries in which they will be sold or implemented.

1.5.2 CODES AND STANDARDS

A quasi combination of professional organizations and government representation creates codes and standards. The following engineering professional organizations are engaged in developing codes and standards in their areas of specialty:

- **American Association for the Advancement of Science (AAAS)** – society to advance science, engineering, and innovation throughout the world for the benefit of all people.

- **American Academy of Environmental Engineers and Scientists (AAES)** – organization of professionals concerned with the environment.
- **American Institute of Chemical Engineers (AIChE)** – organization for chemical engineering professionals.
- **American Institute of Mining, Metallurgical, and Petroleum Engineers (AIME)** – the professional association for mining and metallurgy.
- **The American Ceramics Society (ACERS)** – organization of professionals in the field of ceramics science and engineering.
- **Acoustical Society of America (ASA)** – professionals interested in all branches of acoustics, both theoretical and applied.
- **American Institute of Aeronautics and Astronautics (AIAA)** – organization of professionals in the fields of aeronautics and astronautics.
- **American Nuclear Society (ANS)** – international, scientific, and educational organization that unifies professional activities within the various fields of nuclear science and technology.
- **American Society of Agricultural and Biological Engineers (ASABE)** – professional society devoted to agricultural and biological engineering.
- **American Society of Civil Engineers (ASCE)** – organization of the civil engineering profession worldwide.
- **American Society for Engineering Education (ASEE)** – professional society dedicated to promoting and improving engineering and engineering technology education.
- **American Society of Heating, Refrigerating, and Air-Conditioning Engineers (ASHRAE)** – organization of professionals focused on building systems, energy efficiency, indoor air quality, refrigeration, and sustainability.
- **American Society of Mechanical Engineers (ASME)** – organization of professionals that enables collaboration, knowledge sharing, career enrichment, and skills development across all engineering disciplines.
- **American Society for Testing and Materials (ASTM)** – the organization that creates and maintains standards for testing and materials.
- **BMES (the Biomedical Engineering Society)** – the professional society for students, faculty, researcher, and industry working in the broad area of biomedical engineering.
- **Computing Sciences Accreditation Board, Inc. (CSAB)** – professional organization in the United States, focused on the quality of education in computing disciplines. The Association for Computing Machinery (ACM) and the IEEE Computer Society (IEEE-CS) are member societies of CSAB.
- **International Council on Systems Engineering (INCOSE)** – the professional society in the field of systems engineering.
- **Institution of Engineering and Technology (IET)** – organization to inspire, inform, and influence the global engineering community, supporting technology innovation to meet the needs of society.
- **Institute of Electrical and Electronics Engineers (IEEE)** – the professional association for the advancement of technology.

Engineering Design 15

- **Institute of Industrial and Systems Engineers (IISE)** – the professional society for industrial engineers.
- **Institution of Mechanical Engineers (IMechE)** – professional engineering institution headquartered in London, which focuses on "improving the world through engineering."
- **Institute of Transportation Engineers (ITE)** – educational and scientific association of transportation professionals who are responsible for meeting mobility and safety needs.
- **National Fire Protection Association (NFPA)** – organization for fire protection professionals.
- **National Sanitary Foundation (NSF) International** – organization concerned with protecting and improving global human health.
- **National Society of Professional Engineers (NSPE)** – organization of professionals concerned with professional engineering registration and licensure across all engineering disciplines.
- **International Society for Optical Engineering (SPIE)** – the organization focused on advancing an interdisciplinary approach to the science and application of light.
- **Society of Automotive Engineers (SAE)** – organization of engineers and related technical experts in the aerospace, automotive, and commercial-vehicle industries.
- **Society of Manufacturing Engineers (SME)** – the organization focused on all aspects of manufacturing.
- **The Society of Naval Architects and Marine Engineers (SNAME)** – the professional society that provides a forum for the advancement of the engineering profession as applied to the marine field.
- **The Minerals, Metals Materials Society (TMS)** – the professional organization for materials scientists and engineers that encompasses the entire range of materials and engineering.
- **Society of Photo-Optical Instrumentation Engineers (SPIE)** – the professional society for optics and photonics technology.

Each professional engineering society publishes its codes and standards on its website. Sometimes there is a fee for obtaining a copy of standards documents published by the organization. You may be able to find the standards documents in your university library. As part of your preparation for becoming a practicing engineer, consider joining the professional engineering society for your discipline. It is the best decision you could make. The cost for students is very reasonable, usually in the range of $25–$50. As a bonus, you can list your membership in the professional engineering society in your resume.

Codes are a set of rules that are the generally accepted guidelines or requirements for the industry. Codes are created for safety, quality, and standardization benefits. For example, building codes for a town or city exist to ensure safety, reliability, durability, aesthetics, etc. A code is not a law but can become a requirement enforced by registration and licensing, for example, electrical and plumbing codes for buildings. Codes are intended to be widely applicable as best practices or for compatibility and standardization across the industry. A city, state, or county may adopt appropriate codes, and then, it becomes a requirement to follow the codes. Check your design project attributes for compliance with state, city, town, and county codes.

1.5.3 Specifications

Specifications are typically created by an industry group, consortium, or professional organization. The purpose of specifications is to allow even competing industries to agree to a set of standards that each can use to create products or services and still be compatible. For example, IEEE develops specifications for hardware (and software) developed for internet communication. Specifications may not be regulated or subject to codes and standards. Still, if your design does not comply with appropriate specifications, then it will not be accepted by your potential consumers. Specifications are published and available. An internet search leads to the appropriate one for your design project.

1.6 CAPSTONE DESIGN PROCESS

The capstone design course(s) in the curriculum were included by your engineering program to respond to the ABET requirements for accreditation directly. So, the capstone design course in your engineering curriculum was created to prepare you for engineering practice by placing you in an open-ended, real-world design project with characteristics similar to what you would experience if a company hired you. Your professor(s) guide you through the process that they have developed for that purpose.

However, capstone design is different in two different aspects. The project timeline is first fixed by the start and the end of each academic period (semester or quarter). Some engineering programs have a capstone design course only during one academic period (10–14 weeks). Others have a full academic year experience. The fixed timeline for all design projects creates some unique challenges in managing the capstone design project different from those encountered in the industry.

The second difference is funding and resources. There exist many different models at different institutions for funding capstone design projects. Capstone design projects come from various sources: industry, government labs, faculty research, design competitions, entrepreneurs, or an exercise design problem created for the class. Regardless of the source, they have to conform to the requirements for the capstone design program, which is, at the minimum, those requirements listed in the ABET accreditation criteria document.

This book describes a process I have used for the last 10 years, which has been refined every year and has proven to work well for projects from all the sources mentioned above. The process is broken down into two parts to allow flexibility for engineering programs. Engineering programs that only have one course covering their capstone design course can use Part I. Engineering programs that have a full academic year design experience can use the entire book (Parts I and II). Consequently, the capstone design process timeline maps to the academic calendar, and the requirements are more specific.

Figure 1.4 shows the process flow for Part I of the capstone design. We have broken down the capstone design process into two parts: Part I consists of defining the

Engineering Design

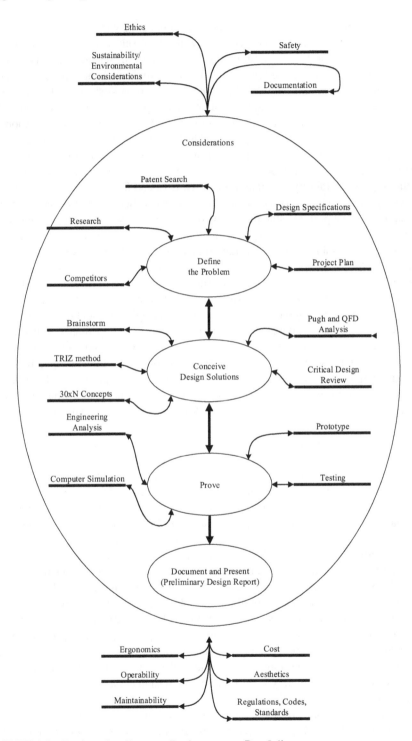

FIGURE 1.4 Engineering Capstone Design process Part I diagram.

problem, conceiving design solution concepts (as many as you can), using decision-making methods and expert and peer review to down-select to a small number of plausible solutions, prototyping/simulating and proving your selected design concept.

Capstone design projects are open-ended design problems. When the sponsor creates a problem statement, they will be asked by your professor who is teaching the capstone to provide an open-ended design problem if possible. Your professor for the capstone design course(s) will assess the problem statement and may ask the sponsor for adjustments to make the problem suitable for capstone in the context of scope, timeline, resource requirements, and level of difficulty. The problem statement will have some realistic constraints such as aesthetics, safety, cost, ergonomics, environmental impact, operability, and maintainability. Cost and financial considerations are always significant factors for capstone design projects. Because the design problem is open-ended, the problem statement provided to you and your team will miss some critical information that your design team will need to research and define. There will be many possible solutions to the design problem, some better than others. Your team will need to follow a rigorous step-by-step (axiomatic) design process, as shown in the capstone design process figure. To define the problem, you will need more information. You can find that additional information in the following ways: researching online information and technical publications, searching through the patent databases, seeking advice from experts, consumer surveys, user surveys, and potential competitors.

Once you have completed the problem definition step, you and your design team can describe and constrain the problem by completing a set of design specifications for your design problem. The design specifications are central to many of the future steps in the design process, so your team must spend the necessary time and intellectual effort in creating a plausible set of design specifications. The specifications should be as complete as you can achieve. If you do not know the numerical values and ranges of some parameters initially, write them down and keep them in your design specifications to be determined (TBD). You will be able to fill those holes later in your design process. It would be best if you began your efforts in design concept generation by meeting as a team and generating as many different but viable design concepts as you can produce during 30–60 minutes of meeting time. Drawings and schematics are effective ways to capture your ideas. Each member of the team should create at least 30 different design concepts. Producing 30 different concepts seems like a daunting task, but you will be surprised by what you can achieve if you push yourself! Share your ideas with your team members. Work together as a team to write down the pros and cons of each concept. You will then use more formal methods to analyze your ideas and select a small number of design concepts to develop further.

A formal part of the process is called the critical design review (CDR), where you present your selected two to four concepts to your peers, experts, sponsors, mentors, and faculty in a group presentation. You will receive a critique of your design concepts and feedback on improving your ideas. With the additional information and feedback you have collected, you will refine your design choices further and create a model or prototype of your design.

You will assess your prototype against your design specifications and document the assessment. You will be required to produce a comprehensive design report for Part I. You will provide the report document to your professor as well as to your sponsors. Ask them for comments and feedback. We call that document the preliminary design report (PDR). The chapters that follow will describe details of each step in the process.

Part II consists of building your design, testing it, redesigning, and repeating until a more optimal design is achieved. In the end, you will document your design and present your work and accomplishments to your peers, professor(s), and sponsors. The building of the design may involve an actual physical build, or it could be a computer design, process model, or software that your team will produce. The building process details will be different, but you will need to prepare a detailed project plan to achieve your build development goals. If you are building a physical device or system, you will need to purchase materials and parts to create your physical design. In most cases, your physical build will not be the model's actual build with the dimensions and materials that will be needed for full functionality. Industry sponsors will generally prefer to build their versions of your design in-house to mitigate any potential liabilities and incorporate your work into what they need.

In your planning, you will have to allow for the administrative work required to process a materials and services purchase request and for availability and shipping time. You will have to consult your professor for the course to follow your engineering department's procedures or the institution has established for purchasing. If you need to purchase software or plug-ins for existing software to create your model or software, you will need to follow the procedures for that type of purchase, which may be different from buying materials. Your team should also pay attention to your skills in manufacturing, modeling, and software development. Do you have the needed skills to build your design? Do you have access to people who can help your team make what you have designed? Do you have the resources that you need, such as funding, tools, software, machines? If you cannot create it, then you may have to redesign it.

Once you have achieved a build of your design or model, you need to verify that it meets your design specifications. You validate your design by creating a set of experiments or tests to test your design against a range of parameters from your design specifications. If there were any gaps in your design specifications, now would be a good time to revisit those and establish values for those. If you still cannot define those parameters, then maybe they should be eliminated from your design specifications. You may need to perform some tests to determine or refine your design specifications' numerical values. You will need to carefully plan your test engineering activities and limit your testing to those that will produce helpful information for your redesign and optimization work. Don't engage in a test because you know how to do that test from your previous courses. Analyze what you need to test and how you would perform that test. Do you need to test for compliance with any regulations or codes concerning your design? Many standard testing methods are published through national professional engineering societies. Make sure you follow those testing procedures if appropriate for your design.

From your test engineering procedures, you will collect data that you can analyze for the performance measurement of your design. The results of the analysis should guide you in optimizing your design or correcting errors. You will have to repeat the cycle of build-test-redesign as many times as necessary to achieve a good design. Many of my past projects have ranged from two iterations to 150 iterations. It depends on the scale and complexity of the design and its build. Figure 1.5 shows the process flow for Part II of the capstone design.

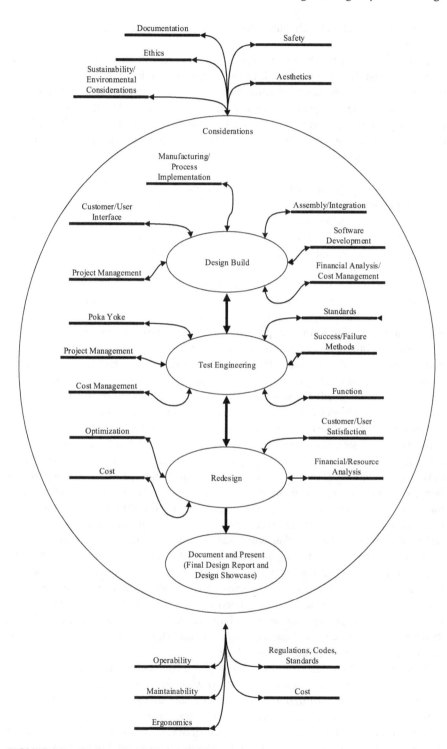

FIGURE 1.5 Engineering Capstone Design process Part II diagram.

Engineering Design

1.7 PREPARING FOR CAPSTONE DESIGN

Capstone design experience is unlike any other lecture or laboratory course in the engineering curricula. The purpose of the engineering capstone design courses is to help you prepare as engineering graduates for working in an engineering or technology industry. The capstone design experience is about practicing the process of engineering design on an open-ended, real-world design problem using the knowledge and skills that you have accumulated in all other courses in the curriculum, internship/coop experience, study abroad, professional engineering society activities, hobbies, and do at-home projects.

The fixed factor in your past experiences is what you have learned in your engineering curriculum before coming into the capstone. Capstone courses typically have several prerequisites to ensure that students take capstone as a culminating experience in their engineering education. So, it would be best if you were at least a senior (4th year and beyond) engineering student when you are in the capstone course(s).

Engineering curricula are structured around five components: science, mathematics, engineering science, discipline-specific engineering, and general education. Figure 1.6 shows the approximate weighting of the number of credits for those

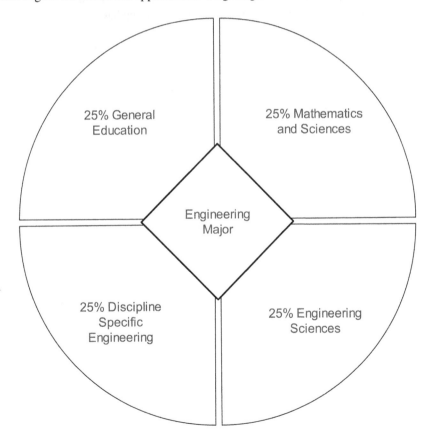

FIGURE 1.6 The average composition of engineering curricula.

components for a semester model. A quarter model would be similar in percentages. Figure 1.6 shows the composition of a typical engineering bachelor of science (B.S.) curriculum in the United States.

Different engineering disciplines may vary, some in the number of engineering sciences (statics, dynamics, the strength of materials, fluids, thermodynamics, etc.), some have more mathematics requirements, some have a computer programming course, etc. You will need all of your knowledge gained in the curriculum to engage in capstone design projects. Your math and science skills provide you with the flexibility to understand and apply concepts directly outside of your field; for example, a mechanical engineering student may be working on a laser project, or an electrical engineering student may be working on an electronics cooling project. The engineering sciences and discipline-specific courses prepare you for design modeling and engineering analysis. Laboratory courses prepare you for designing and conducting experiments related to your project. The general education courses prepare you for critical thinking and analysis of the aspects of the project dealing with environmental impact, ethical issues, political considerations, safety/risk, financial analysis, project management, aesthetics evaluation, etc. In capstone design and this book, we emphasize, review, and introduce the knowledge that may not be covered in many engineering curricula before capstone. You may need to create or update your resume, which your professor and the sponsor may use to select you for a particular project based on the skills, experience, and knowledge that you have listed in your resume. I use the information in my student's resumes to assign them to particular projects.

2 Design Project Team

During the past three decades, there has been a profound change in the practices in the industry in performing engineering design work. Many years ago, the larger engineering companies practiced technical groupings of engineers in design divisions such that all mechanical engineers worked in one group, and all electrical engineers worked in another group, etc. Gradually, that practice changed as companies discovered that diversity of backgrounds and expertise produced the better project and design outcomes. They also learned that it was best to have teams of people from all educational backgrounds that can produce more robust design results. The technically excellent engineers who graduated from elite engineering programs but worked in a lone-ranger way proved to be less effective than teams of engineers who could work together effectively.

In capstone design, similar to industry, teams must exhibit a cooperative attitude and work together in unison. In industry, people are assigned to projects based on their skills, qualifications, and experience. Frequently, people assigned the same team have never worked together before and must develop working relationships that effectively advance the overall team towards the project's goals.

Members of the team should become focused on completing the design project and depend on each other in performing the tasks. Capstone projects are complex, multifaceted design problems and require diverse backgrounds and experiences for best results. Team members should define the roles they will play to conduct the design project; for example, design engineer, test engineer, team leader, research engineer, financial analyst, sales engineer.

The capstone team will manage its work by planning and communicating with sponsors and the professor. Team members may be all from one discipline of engineering, or they may be from two or more disciplines, with the majority of the team members most likely being engineers. Different engineering programs and colleges practice the diversity of disciplines in their particular way: some programs have only their own majors on the team; others may be assigned engineers from different disciplines. Regardless of the disciplinary composition of the team, the diversity of backgrounds among team members is essential to the success of the project.

Each capstone team will determine the best methods to communicate among themselves and hold meetings. However, a regular (typically weekly) progress report will be an excellent tool to update the sponsor and the professor on the team's work status. The weekly progress report content and format will be covered in a later chapter.

Teams should remain together and stable throughout the design project duration. However, there may be instances where a team member is not contributing to the project. The team members should counsel members who are not engaging and participating. If team counseling is not successful, the professor for the class may intervene to investigate the problem and take appropriate corrective action.

Just because a team has been formed and has the attributes mentioned above, it does not guarantee that the design project will be successful. There is much more to the success of capstone design projects, as we will cover going forward.

2.1 SELECTION OF TEAM MEMBERS FOR DESIGN PROJECTS

Success capstone design projects result from following a structured approach as well as who is assigned to the project team. A foundational goal for capstone design is to prepare students for professional engineering practice through the simulated environment created in the capstone courses. In industry, assignments to teams are based on skills and qualifications best suited for the project. Following a similar process in the capstone experience will result in a more capable team of students and better outcomes for the design project. There are several different ways to assign students to capstone design project teams. The challenge for the professor and the students is to match a team of students for each of the projects identified for the capstone course from various sources. Typically, some N number of projects are provided or determined by the professor teaching the capstone class. Several different approaches can be taken in assigning students to the teams.

One approach is to assign students randomly to each project at the start of the year. Each project will have a team, and all students will be assigned to some projects. The results of this approach will also be randomized as to the success or failure of the project teams. Generally, this is not a good approach because there are better ways.

Another approach is to present the project list for the year and have students express some level of interest in each project. Assign students based solely on their level of interest in each project. The problem with this approach is that students migrate towards working with only other students whom they know or are friends with. People do not normally choose their team members in a suitable work environment, but team members are assigned to the team by a supervisor. The situation also presents a challenge to the professor who needs to assign student teams for all projects and ensure that every student is assigned to an appropriate project. Like the musical chairs game, some students will be left standing!

Another approach is to have the student team select their own capstone design project, probably with the professor's approval for the class. The professor will review such projects carefully to ensure an equitable level of complexity and difficulty. Resources for conducting such a project may become an issue depending on the particular administrative policy of the engineering program or the institution.

An approach that has worked well through experimentation with many projects over the years is to assign students to project teams based on their qualifications. I have tried all of the above strategies, and this one works best and has produced the best outcomes. In this approach, the team assignment is similar to the industry because students are "hired" into the teams based on a statement of qualifications, academic performance, internship or coop experience, diversity, and a resume summarizing their experience. Using capstone design as a "hiring" experience also helps engineering seniors who will be seeking jobs during the final year of study in preparation for graduation. It provides the students with the opportunity to prepare/update their resumes, which will help them apply for jobs and interviewing.

Design Project Team

Regardless of what approach is used to assign students to design projects, each student in the class must be assigned to a project team very quickly at the start of the academic year.

2.1.1 Background Research on the Project

Design project sources can vary depending on the approach and philosophy taken within the engineering program, department, and college. Most engineering programs desire a close relationship with the employers of their students, so they look for design project opportunities with those companies.

As a student, you probably desire to be on specific projects with those companies in which perhaps your friends or family members work. As soon as you learn about the design projects from your professor or company presentation of projects, you should thoroughly research the background for those problems of interest to you. If you are asked to write a statement of qualifications, you can summarize the results of your background research on the problem along with your credentials. Your interest and ability to conduct the background research successfully are an asset to the project team that will positively assign you to that team.

You can find background information on the project through several direct means. First, talk with your professor about the project. Learn as much as you can from what the professor knows about the project. Take notes, so you do not forget the information shared with you. You can use your notes to do further research later.

Second, contact the sponsor for the project and ask them as many questions as you can think of, and again, take notes. The sponsor is perhaps the most knowledgeable person on the project. They can share more details with you than what they presented to the class or point you to others in their company who know the problem. Third, you can research through the patent database if the project is a product that may have similar or related inventions. You will have to think broadly about the design problem to find information that may be useful. If the company has a patent related to the project they are sponsoring, look for that patent. Also, look for other patents that the company or the sponsor may have filed. Fourth, look for scholarly works that may have more information about the design project. Look for open ware if it is a software project.

Google Scholar (scholar.google.com) is an excellent free source. You can also search in ResearchGate (researchgate.com) and Academia (Academia.edu). Another source of scholarly publications is your school/university library search database. Fifth, look for competitor products/processes online. Ask the sponsor who their competitors are and use that as a starting point. Be flexible in what you consider to be an exciting project. This effort is highly worthwhile and will positively influence being considered for an assignment to a project of interest to you. Interest in a broad range of technical areas will also serve you well later in your professional career.

2.1.2 Educational Preparation

Capstone design sequences are scheduled as a culminating experience for engineering students during their last year of study towards a Bachelor of Science degree in an appropriate engineering discipline. The capstone design experience is meant to be a culminating experience utilizing all knowledge learned in earlier coursework in an engineering curriculum. The four components of the engineering curriculum are mathematics, sciences, engineering sciences, discipline-specific engineering courses, and general education. Engineering professional electives may play a role if the topic of the professional elective course aligns with the project.

Because all students in one major take the same set of required science, math, and engineering courses, the only distinguishing differences among students' educational experience is in general education or if students are also pursuing a minor, a second degree, transfer from another major, or simply taking additional courses. If the team consists of only students in the major, less educational diversity exists among the students. If students are from two or more disciplines, their diversity of knowledge will help the team with multidisciplinary projects.

Students can choose professional electives that are better aligned with their capstone project and be better prepared for the project.

2.1.3 Professional Preparation

Capstone design experience is a form of practice and preparation for becoming an engineering professional. However, professional conduct is a necessary part of being assigned to a capstone design team. Sponsors of capstone design typically look for prospective future engineers to hire from the design team. The sponsors also expect team members qualified for the project and can positively and professionally contribute to the design solution.

All engineering students are required to successfully complete a set of courses outlined in the curriculum for the major. So, all students coming onto the capstone design courses have a minimum educational background. Professional elective course in the engineering major is one form of students distinguishing themselves in particular specializations which can be helpful to the appropriate design problems.

Many engineering students engage in internships or coops. Some engineering programs require a coop on internship. The experience of working as a coop student or an intern is a powerful way that students can prepare professionally before entering capstone design. The internship or coop may have been with the project's sponsor, which will help facilitate contact, communication, and gathering information on the design problem.

Any prior or concurrent industry work experience will benefit the engineering students in working on real-world design problems and creating design solutions. The enrichment that is provided by working outside the academic boundaries is invaluable to capstone design success.

2.1.4 Qualifications

Students should apply for and be assigned to capstone projects based on their qualifications appropriate for the project. Different capstone design projects have unique needs for technical and practical expertise and experiences. Student qualifications that should be considered and expressed in the application for a capstone design project include:

- Academic performance in required engineering, science, mathematics, and other courses; a transcript is a good indicator of this performance.
- Professional and other elective courses relevant to the requirements of the design problem
- Internship and coop experience
- Experience working in a research lab
- Experience working in a team (could be athletic, professional, military, or other)
- Any special skills or certifications, especially those relevant to the design problem
- Hobbies, extracurricular, or extramural interests that are relevant
- The explanation of why you are interested in the project and why you should be assigned to it

In summary, treat the application for a design project as practice for applying for a job. List and explain all of your qualifications and explain why you should be selected to work on the particular project. Remember that many other students will be applying for the same project, so your application should be better than others to be selected.

2.1.5 Professionalism in Interactions

Engineers are expected to be truthful, tactful, reliable, respectful of others' opinions, and effective communicators. Engineering students in capstone design are preparing for their profession after graduation. Therefore, they are expected to behave professionally in the class and their interactions with sponsors, mentors, advisors, teaching assistants, and professors.

Professionalism includes self-presentation (how one dresses), attitude, and the way one communicates. Etiquette, respect, and thoughtfulness are essential qualities to professionalism.

Employers value employees who professionally fulfill their duties. Employees with a high degree of professionalism are deemed more reliable and credible by their employers. Professionalism and etiquette can give an advantage to students who want to be assigned to their top choice project. Later in life and in the profession, professionalism can advance the careers of any engineer.

Professionalism is a combination of many traits. We will discuss some of those in what follows.

Professionals pay attention to details, sponsor needs, requirements, and communications. They are responsive to clients, sponsors, mentors, and fellow team members. Professionalism includes being proactive in anticipating issues, problems, and actions needed to solve problems and issues. Being proactive requires that the professional is interested in the project and pays attention to details.

Professionals are considered to be reliable and dependable individuals because they honor their commitments. Commitment shows responsibility to the assigned duties and tasks. Being punctual and prepared for scheduled meetings is another way of demonstrating commitment and interest. Commitment also is an expression of respect for team members and stakeholders in the project.

Professionals welcome the obligation to their project goals and see themselves as team members with total commitment to the project's success. They are considered to be respected and valued team members by project stakeholders.

Professionals dress according to the norms of their work environment. They are respectful of others and observe an appropriate etiquette in interacting with fellow team members and project stakeholders. They keep interference from personal matters to a minimum and mitigate issues affecting team members' productivity and efficacy on the project.

Professionals strictly observe the policies established by the sponsor concerning confidentiality and non-disclosure of project information. Many sponsors require that project team members sign a non-disclosure agreement (NDA). Team members on the project should read the NDA carefully and thoroughly and make sure that they understand and agree to the terms and conditions listed in the NDA document. Once they have signed the NDA, they must honor their commitment to the NDA terms for the duration of the agreement.

2.1.6 Team Dynamics

B.W. Tuckman and Jensen (1977) suggested a four-stage developmental process for project teams to transform into productive teams as Forming, Storming, Norming, and Performing.

Forming teams takes time. When a grouping of three to five students is assigned to the work together, they are likely not to know each other well. They could be strangers in an extensive engineering program. The group must recognize that they must transform from the initial grouping into a unified team working on the same goals and objectives. Students are excited to be on a project team when the group is formed. They have a positive attitude and are polite. Some are anxious because they do not understand what work the team will do. Occasionally, a student may be disgruntled because the team they were assigned was not their first choice. This stage can last for a while until the students start to work together and make an effort to get to know each other and their assigned project.

Next, the design team starts marching forward, where team members begin pushing the boundaries established during the forming phase, entering a storming phase. The storming period can become the start of a path to failure for some teams or team members. Conflicts among team members may emerge and can cause frustration among team members. Because there are many unknowns about the design problem and the solution is not clear, students can become overwhelmed by the amount of work, frustrated with the requirements of the class, unable to accept the structured approach, and feel lost. Some students may question their qualifications to do the design project and the worth of the whole effort. Students who stay on task may experience stress, particularly if they do not have strong relationships with their peers or if they cannot solely rely on what they learned and practiced in their earlier coursework.

The design team will gradually transform from the storming to the norming phase. Students start to work through their differences during the norming stage, and the scope of the design problem starts to sink in. They develop respect for each other and authority. The team members begin to socialize, and they develop the ability to ask each other for help. The group starts working together more like a team. The team begins to make progress towards their design project. The team may relapse into the storming stage as new challenges emerge. The team starts performing when the hard work of team members starts producing results. Students get positive feedback in response to their work, such as successfully defining the design problem, creating credible design specifications, and finding valuable and applicable information from their search. The structure and process of the capstone design become rewarding for the students. They feel good and relaxed about being a member of the team.

The last phase is the adjourning stage. All capstone design teams adjourn at the end of the course(s). In many engineering programs, the conclusion of the capstone design experience also coincides with graduation. The knowledge gained by the students during the capstone design builds a bridge to their jobs, careers, and possibly graduate school.

Capstone design teams are self-managing, self-organizing, and are centered around member leaders. There is no top-down management on capstone design teams. Member leaders coach each other, remove obstacles, and mitigate distractions for the team to flourish. The whole team is measured for their performance. Individual contributions are aggregated into the overall team's performance assessment. Each member of the team will do whatever jobs are needed to ensure the success of the team. Teams should have three to five members for the best utilization of the human talent on the team.

Members of the capstone design team should continuously learn from each other and teach one another new skills and knowledge. Members demonstrate a commitment to each other and the project. Team members strive to reach agreement and consensus in their decision-making. Each team member works to remove bottlenecks and is willing to work on different tasks as needed by the team and the project. Team members facilitate communication and build relationships with their sponsors and mentors.

When there is dysfunction on the team, team members work collaboratively to resolve the issues and re-engage team members who experience distractions and low productivity.

2.1.6.1 Effective Team Membership

A team is a collaborative effort. Working in a team is not straightforward. You will have agreements and disagreements with your team members during your interactions and discussions. It is paramount not to lose sight of the team's goal to create a successful design solution.

Being productive team members requires some core qualities that team members must have or develop during the capstone design project. The design team membership qualities are a positive attitude, a problem-solver, curiosity, dedication, creativity, responsiveness, and appreciation of diversity.

Positive Attitude – attitude is what makes or breaks design projects. Having a positive attitude towards the project and your team members is towards a successful design project.

Being a Problem-solver – as an engineering student, you have learned and been educated in solving engineering problems. Those skills will be needed for the technical parts of your design project. You also need to be a problem-solver for all other aspects of your design project, working with your design team, administrative challenges, interaction with your sponsor, and working through all of the capstone course requirements.

Curiosity – capstone design projects are typically complex problems. There are no shortcuts or quick and easy answers. Curiosity about the design problem is essential to work through the many steps needed to find a plausible design solution.

Dedication – open-ended design problems require particular persistence and commitment on the part of the team members. Commitment to solving the design problem and to fellow team members is an essential quality for success.

Creativity – design problems provided to capstone are typical problems that have not been solved before or require a new and creative solution. Industry sponsors of capstone design projects value the creativity of young engineering minds the most.

Responsiveness – the timeline for capstone projects is fixed by the academic calendar. There is no time to waste. You must be responsive to team members, sponsors, and others in all of your communications.

Appreciation of Diversity – the ability to listen, understand, respond, and adapt to the various perspectives and backgrounds of team members. Working in a diverse technical and cultural environment is an essential quality of being a productive team member.

2.1.6.2 Team Leadership

In a capstone design team, one person should take the role of the team leader. Their role will be to schedule the team meetings, handle external communications for the team, and help the team maintain a project plan and turn in reports and assignments before deadlines. The team leader must be an effective communicator and handle all communications in a timely manner. The team leader will advocate for the team and plan and facilitate internal team meetings and meetings with sponsors and the professor. The team will elect a team leader, or the course professor may assign one.

2.1.7 ROLES AND RESPONSIBILITIES

Teams should structure themselves in a way to best achieve their purpose. In collaboration with their fellow team members, individual team members can define their areas of expertise or concentration on the team, for example, design engineer, research engineer, team leader and manager, financial analyst, and marketing engineer. Be creative to define your title and role on the team.

2.1.8 EFFECTIVE TEAM MANAGEMENT

Team meeting time is precious, and you do not want to waste time during the meetings. Time is the most valuable resource for the teams. For productive team meetings, it is essential to have a regular schedule (one or two times a week), an agenda relevant to all team members and the project, and minutes of the meetings that show essential discussions were held, and key questions were considered and answered.

Design Project Team

2.1.8.1 Scheduling

Many engineering seniors have part-time jobs while attending the university, making it more challenging to schedule a standard time. All members of the team must allocate the time for preferably two team meetings per week. Typically, capstone design courses are expected to allow common periods for students on a team to meet and do their design work.

Teams should schedule their regular and other meetings on their project plans, so the time is set aside as part of the required project and class activities.

2.1.8.2 Agenda and Time Management

Agenda for team meetings should be prepared by one member of the team, usually the team leader, and shared with the team members before the meeting. Members should be invited to submit agenda items to the team leader for inclusion if time permits. Those items can be considered either at the immediately scheduled meeting or at future meetings.

To engage all team members to participate in the meeting discussion, pose agenda items as questions that team members can answer. Table 2.1 shows an example table that you can use to create your agenda table. Ensure that each team member has a precise and concise topic with time duration for that topic. Strictly adhere to the agenda and time duration for each topic. Allow enough time for each topic to make sufficient progress during the meeting. Be careful; it is easy to underestimate the amount of time for the item.

Include any documents or handouts needed for preparation with the meeting agenda and send out the agenda in a sufficient amount of time for all team members and allow enough time for them to review that material before the meeting.

TABLE 2.1
Creating a Practical Meeting Agenda

Topic	Duration	Purpose	Lead	Preparation	Process
What do we need to add or change in our project plan?	10 min	Decision	Sarrah	Review the project plan	Sarrah presents some options
What questions can we ask about the problem so we can define it better?	15 min	Decision	Josh	Read problem definition	Josh presents some ideas and questions
What design specifications should we include?	15 min	Decision	Austin	Review design specifications development template	Austin will guide the team through the template questions
Was this a productive meeting? What should we do better for the next meeting?	5 min	Discussion	Hope	None	Team members identify pluses and minuses for this meeting. Agree to make changes to correct minuses

Allow time at the end of each meeting to review the effectiveness of this meeting and how you can improve in your sessions in the future. Table 2.1 shows an example agenda that you can modify for your meetings.

2.1.8.3 Minutes

Minutes of meetings are an official record of what was discussed and decided during a meeting. Minutes allow team members and others (professor, sponsor, etc.) to review and evaluate the team's progress or note exceptional accomplishments or deficiencies. The minutes of your meetings should include the following information:

- Date, time, and duration of the meeting
- Location of the meeting
- List of names and e-mails of who attended the meeting
- The name of the team
- The title of the project
- The sponsor of the project
- A listing of the agenda items and minutes for each item of the agenda

Minutes should be short (no more than one page per meeting). You can attach any supporting documents, such as the project plan to the document (or a link to it).

2.1.8.4 Action Items

Action items are a list of tasks/actions that the team needs to perform. The list will grow and shrink as new actions are identified or ones on the list completed. The action items should be added to the project plan and be tracked for time and responsibility.

Each action should be planned carefully to ensure relevance to the project's progress and success. A list of action items is similar to a to-do list. A team should be able to present this list when asked and discuss the rationale behind each action item on the list.

2.1.9 PROGRESS REPORTS

Progress reporting is an expected activity in engineering projects. The purpose of the regular progress reporting is to keep everyone informed of the project's status and progress towards the goals and objectives. Progress reports should also be used to update the project plan.

Progress reports make monitoring of engineering design projects possible. In industry, project costs have to be monitored carefully to mitigate cost overruns on projects. Customers expect the best product for the lowest price. So, progress reports must include assessing the impact of the team member's activities on the project timeline.

A useful format for a progress report is a memorandum from the team to their supervisors, i.e., professors and sponsors. The active time duration for progress reports for capstone design projects is on a weekly schedule. On a designated weekly deadline, the team will be required to submit their progress report, usually by e-mail, to their sponsor and capstone professor.

Design Project Team 33

The progress report template is provided in the electronic materials for this book. The report should contain the following information as a minimum:

- Date of progress report
- Names of individuals the progress report is addressed to (sponsor, professor, etc.)
- Title of the project sponsor of the project
- Duration covered by the progress report
- Names of students (and their project role/title) who participated in preparing the progress report
- A short recap of the project
- Description of progress made or the results accomplished during the past week detailed by activities of each team member
- Description of any problems or issues that may have surfaced
- Action plan by the team to resolve problems or obstacles
- Description of what is planned for the next reporting period
- Attach an updated project plan (Gantt chart)
- Attach any other relevant documents such as drawings and articles.

During the project analysis that will take place for each progress report, the team should examine the tasks in progress and look for opportunities to achieve parallelism. If possible, reduce the number of any tasks that had been scheduled sequentially.

2.1.10 Communications

Effective communication is an essential skill for success in life, but it is critical for capstone design. The members of an engineering design team must communicate effectively within the team and with all of their project stakeholders. Current methods or venues for capstone design communication include:

- Verbal/oral face to face meeting
- Verbal/oral by phone or software (e.g., FaceTime, Skype, Duo, WebEx, Zoom)
- Videoconference (e.g., Skype, Duo, WebEx, Zoom)
- E-mail

The method of communication that you will use for your project will depend on what software and facilities are available to you at your university and the sponsor's preferences for the method of communication. Some sponsors may have confidentiality and security concerns, which will require the use of secure certified software for communication. You will need to check with your sponsor regarding any security requirements for your interaction with them.

Figure 2.1 shows a mapping of the different methods of communication. For each method of communication, the clarity of communication is mapped to the information content. A face-to-face meeting is an effective communication method because of the rich nonverbal content that enhances the context of the meeting. It would be

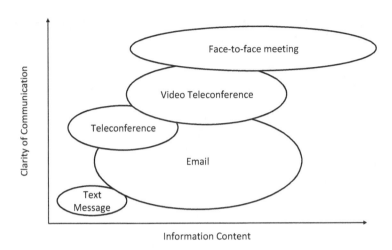

FIGURE 2.1 Comparison of different communication methods.

TABLE 2.2
E-mail Strengths and Weaknesses

Strengths	Weaknesses
Quick creation	Too quick!
Fast delivery	Fast delivery
Location independent	Lacks surround information (e.g., eye contact, body language)
Audit trail of the communication	A lengthy message with audit trails of everything
A facility for providing feedback	E-mail is sent, but is it read?
Allows for attachments	Attachments
Environmentally friendly because it reduces the use of paper	Lower quality communication
	The sender assumes that the recipient will take action
	Low security and confidentiality

best if you analyzed the effectiveness of your way of communication in the context of the composition of the personalities in your team and the possibility of misunderstanding. If your message is misunderstood, how easy will it be for you to correct the situation? What are the consequences of miscommunication?

Many of your messages will be sent and received by e-mail. Table 2.2 shows some of the strengths and weaknesses of e-mail communication. Consider the context of your communication and the strengths and weaknesses as you decide to use e-mail. Many students and professionals do not use e-mail properly. The anatomy of an e-mail consists of:

The Address Field, Including" to:" and" cc:" – The e-mail should be addressed to a correct list of individuals who should receive it—the carbon copy field (cc:) should include individuals who should receive a copy. For example, when one team member communicates with the sponsor or the professor, they should copy all team members to be informed of what was said.

Subject Line – The subject lines must be informative and allow the recipient to identify (and later search) the e-mail quickly. Because there are many students (and therefore teams and design projects) in a capstone class, the subject should include information about the team number/sponsor/and project title.

Greetings – Always use a professional greeting, including the individual's title or salutation (e.g., Ms., Mr., Miss., Dr., President, Vice President). Your e-mail communication on the design project is an official message and must have the appearance and the content of a professional e-mail.

Content – The content of your e-mail must be thoughtful and convey your message in a clear and well-organized manner. Your message should be short and not too lengthy.

Signature – Your e-mail message must be signed with contact information such as e-mail and phone number so the recipient can contact you. Your signature should include your team information and the title of your project.

Attachments – Use attachments to include important information that should be shared with the recipient. Explain the content of the attachments in the main e-mail, so the recipient knows what is inside the documents before they attempt to open them. Each attached file should be named such that it can be easily identified by its name when downloaded and stored in a file storage system. It should be easily searchable for later references.

Using e-mail properly can help you and your project and will serve you well in the long run.

2.1.11 Online Communication

As I am writing this book, we are experiencing a world pandemic caused by the COVID-19 virus. The pandemic situation requires social distancing and a move to using online communications technologies for many aspects of the engineering profession and education. The move to use of online technologies for conducting education and business has been rapid but will probably change our working and learning habits for the future.

Capstone design projects have also moved online, more so than previously. Computer files created or found through online search must be stored so that all team members can access them, and each member can add to the file sets for each part of the project. The sponsor may also want access to some of the files that the team is creating, e.g., CAD files or computer codes. Many options are available to the team for storing and managing their common files and data sets.

Teams must also be able to meet remotely to observe the requirements for social distancing during the pandemic. Many options are available to the student teams for software tools for video conferencing. Still, also many companies are moving to remote online meetings for the future.

We will cover some of the more common options in the following two subsections.

2.1.11.1 Online File Management

Storing, retrieving, and syncing computer files stored by the design team across many devices (e.g., desktop computers, laptop computers, tablets, and smartphones) are necessary for a design team operating with members at remote locations. Another essential function is to search the content stored by the team to find items quickly and accurately. Many file-sharing services and tools are available and work well for design teams. We will discuss some of them.

Microsoft Office 365 is available at most universities for students and faculty. Many companies also use Microsoft Office 365. This can be an excellent way to share many files through Microsoft cloud services, including files storage and sharing. Microsoft's Word, Excel, PowerPoint, One Note, and Outlook applications are widely used. The Microsoft Office applications integrate with Microsoft's other services: OneDrive, SharePoint, Skype, and Exchange.

Google's G Suite is an excellent alternative. G Suite includes Gmail, Docs, Sheets, Forms, Slides, Sites, Drive, Calendar, and Hangouts. The Google software runs across all devices capable of running the Google Chrome browser or the Google Chrome operating system. Student team members can easily collaborate on their documents and files within the Google environment.

Box (www.box.com) is easy-to-use secure file storage, file-sharing, and collaboration tool run on Windows PCs, Macs, iOS mobile devices, and Android mobile devices. Dropbox (www.dropbox.com) is another accessible file storage, sharing, and collaboration tool. Google Drive (www.google.com/drive) is available at many universities, or you can also gain access through a personal Google account. Microsoft offers OneDrive (onedrive.live.com) for file storage and sharing, integrating with their suite of office software.

2.1.11.2 Real-Time Conferencing Tools

Meetings online can be held by phone, but video-teleconferencing offers a much more productive environment for online meetings. Video-conferencing tools offer screen sharing, where PowerPoint-style presentations can be shared. Some video-conferencing tools also allow for session sharing, where a participant in the meeting can receive permission from another participant to control their keyboard and mouse. Several options may be available to you through your university.

WebEx is a video-conferencing environment offered through Cisco. WebEx is a secure environment and is acceptable by government laboratories, industry, medical professionals, or any meeting where a safe and secure communication environment is needed to discuss proprietary or company sensitive information. Cisco also offers additional tools such as the Cisco WebEx Event Center, where you can schedule meetings with automated notifications sent to the participants you list.

Zoom (http://zoom.us) is a popular and easy-to-use video conferencing system. During the COVID-19 pandemic, where many people worked and studied from home, Zoom gained many users. Many universities adopted Zoom for their online teaching and learning tool.

Skype (www.skype.com) is a video-conferencing tool offered through Microsoft. It integrates well with the MS Office environment and is easy to use.

Google Hangouts (gsuite.google.com) is a video-conferencing tool offered through Google. It integrates well with Google's G-Suite environment and is also easy to use. G-Suite includes a Gmail account for users. Your university may have implemented the G-Suite such that your account is of the form your – name@myuniversity.edu.

There are many other freeware-type video-conferencing systems available. There will probably be many more new systems that will be developed and available soon because of the significant move towards working and studying from home.

3 Project Management

Capstone design problems are open-ended, complex, and require multiple people and other resources to accomplish. A structured method is needed to manage all of the individual and group tasks that must be performed to complete the project within the capstone course's fixed and rigid time frame and within the budget or cost guidelines provided. The method that is commonly used in engineering and the industry at large is called project management.

Managing a project requires a plan or a roadmap. The plan is specific to the team and the project. It is used to coordinate project resources and activities by the team. It is also used to communicate with others (sponsors, professors, mentors, etc.) about the project's status and progress. The project plan shows the distribution of responsibilities and resources.

Project management is the process of decomposing, estimating, planning, organizing, and managing tasks and resources to accomplish a defined objective within constraints and resources. The resources in consideration include people, facilities, equipment, money, and time.

The purposes of project management include:

- Create a realistic and practical plan.
- Motivate the project team by creating a road map to the end goal of a successful design solution.
- Communicate the design solution process to the sponsors and other stakeholders.
- Foster collaboration among team members by creating a clear scope and statement of the objectives of the design project.
- Align student team members to the common goals of the project and how to achieve them.
- Develop confidence among the team members and stakeholders in team capabilities and achieve a successful design solution.
- See and communicate when things have gone wrong and the team gets off track in their progress.
- Allow sponsors and mentors to see the path of the team and advise or help the team.
- Identify the need for special tools or resources that will be needed and allow time to locate or acquire those resources. If it is not feasible to find those unique resources, then the team will need to change its methodology and plan.
- Allow the team to identify potential bottlenecks or problem areas in their approach – plan workarounds when problems are identified.

Project management starts with a problem decomposition by estimating project details. The challenge for capstone design teams is their lack of experience in project management and a design problem that probably has never been solved before. So, the team does not have all of the information that they need to initially layout the entire project with all of the necessary steps. However, the process we describe in this book is clearly defined. Following the capstone process, the team can create a first project plan and then fill in the details to learn more about their project and how to complete it. Planning should start with a table listing all the team's tasks, the resources needed, time duration, and cost for each step. Table 3.1 shows an example list for a capstone project.

TABLE 3.1
Capstone First List of Activities or Tasks

Task	Estimated Duration	Resources	Dependencies
Organize team meetings	1 day	Team members	
Meet with sponsor	1 day	Team members and sponsor	
Review problem definition and prepare a list of questions	1 week	Team members, sponsors, mentors, experts	
Search for additional information about the design problem, patent search, article search	2 weeks	Team members	Problem definition
Create design specifications	2 weeks	Team members, sponsors, mentors, experts	Problem definition, information search
Generate design concepts	1 week	Team members	Problem definition, information search, design specifications
Analyze concepts and down-select to a smaller number	1 week	Team members	Information search, competition research
Prepare presentation for critical design review	1 week	Team members	Concept analysis
Critical design review	1 day	Team members, class, sponsors, mentors, professor	All previous tasks
Use feedback from critical design review to decide on a concept to prototype	1 week	Team members	Critical design review
Prove design concept – create a prototype/model of the design solution	4 weeks	Team members	
Document the design work – create preliminary design report	2 weeks	Team members	
Conclude Part I by presenting/meeting with the sponsor	1 week	Team members	

Project Management 41

3.1 PLANNING

Project planning is a necessary group activity for capstone projects. During preparation, the team must work together to prepare lists of tasks, the order of events, resources needed, and time duration. Because most engineering students have not had the educational or professional experience with project planning, many of the initial efforts in planning need to be estimated. The estimations can be refined later as the team learns about the project and how they work together and individually towards accomplishing the common goal of a successful design solution. Planning is a continuous process throughout the design project. The result of planning is a project plan. The team will use the project plan to monitor their progress on the design project and communicate with sponsors and the course professor. Each item in the project plan will become a separate task. Each task will need to be specified with the following attributes:

- A descriptive task title.
- Who is responsible for performing the task?
- What are the non-people resources needed?
- When can the task start?
- On what other tasks does this task depend?
- How long will it take?

A reverse planning method works best in capstone design because all capstone projects have a strict end date that maps to the university's academic calendar. So, the team can start with the end in mind and work backward from the last task, which will be a successful design solution and documentation as deliverables. By starting from the end in mind, you can plan your project to achieve that goal within the allotted time. There are usually other check-points along the way that are established by the requirements of the capstone design course (consistent with the needs of the design projects), such as the critical design review presentations and other class assignments.

Those tasks in the project plan become fixed points in time with a specific output. Other duties will be performed during the intervals of time between the fixed points in the project plan or parallel to the required assignments. Planning requires critical thinking about the problem-solving approach in engineering design. The method in capstone design is different than any other problem-solving approach learned in earlier classes. The student designers will need to consider the problem statement and the many sponsor requirements carefully and simultaneously.

Additionally, there are the requirements for the capstone course, the fixed time allowed for the project, and usually minimal funding for the project. In some ways, a capstone design project is more challenging than what the engineers in the industry may face. However, the experience is unparalleled, and the benefit to the students and the engineering curriculum is substantial.

3.2 SCHEDULING

Scheduling is the process of mapping the tasks identified during planning to a timeline. Each task will have a start date, end date, and duration. For some tasks, the start date may depend on some previous task(s), i.e., the current task cannot start until the preceding tasks have been completed. Each task will also have resources associated with it. Resources may be assigned on a percentage scale to a task, e.g., a student assigned to a task may be working on that task only for 50% of their time. A particular resource may only be needed for a small amount of time compared with the duration of the task, e.g., a tool needed for machining a specific part may only be needed for a few hours, while the broader task of manufacturing the design may be scheduled for much longer.

Many project management software tools are available for assisting the engineering design team in planning, scheduling, monitoring, updating, and organizing their project. The scheduling process should be prepared first before using a software tool to capture the information and manipulate the overall project plan and schedule.

Scheduling helps the team to create plausible predictions of time requirements for the project, including problem definition, scope, statement of goals, and what will be achieved. The schedule will help the team to produce cost estimates and financial analyses for the project. A proposal will be created and delivered to the sponsor for review, approval, or negotiation in real-world practice. After the sponsor approves the project proposal based on time and cost estimates provided by the design team, a contract is issued to the design team to proceed. The team will start their project only after the sponsor's approval and schedule (and the professor).

In creating a schedule for the project, the team will have to achieve the following:

- Assign equitable tasks to individual team members.
- Assign appropriate tasks to team members matching their knowledge and skills.
- Maximize the use of people resources on the team to achieve the design solution.
- Account for external dependencies such as purchasing, shipping time, availability, and resource conflicts.
- Include task dependencies and sequencing of activities on the project.
- Load-level all resources such as people, tools, machines, computers, and databases.
- Include the fixed dates associated with the capstone design course requirements. Avoid unreasonable or unjustifiable assumptions about the project and team members' technical skills.
- Avoid underestimating the level of effort that will be required for some tasks, e.g., coding, machining, welding, 3D printing, learning new software.
- Avoid unrealistic predictions.
- Include the best estimate of the probability of completion and success for each task.

Defining that the tasks can be achieved after the team decomposes the overall complex design problem. Problem decomposition is a process of breaking up the more significant problem into more minor problems that we know how to solve. The decomposition method is sometimes called the divide-and-conquer method.

A quick internet search for "project management software" will return many results for options. The choice of professional project management software is dependent on your discipline. Many engineering companies today use Microsoft (MS) Project.

Project Management   43

Project management software is built into some larger CAD systems such as CATIA or is integrated through a standalone software connected to the CAD system, such as PRIMAVERA linked to AutoCAD. Project management software that is cloud-based allows each member of the engineering design team to contribute to and track their activities and tasks in the overall project. Therefore, cloud-based project management software is much more beneficial by engaging team members in the project's success.

3.3 ENTERING PROJECT INFORMATION INTO MS PROJECT

We will use MS Project for examples in this book. MS Project is available as a cloud-based product, or it can be run standalone. The cloud-based version is available through MS, or it can be installed by an enterprise for shared access by team members.

For example, we can start entering our project information into MS Project after completing our work on creating project plans and organizing our notes and data for the project.

Figure 3.1 shows the blank entry field when you open a new MS Project file. The information that you will need to have previously prepared to enter is mentioned below:

- Descriptive task name
- Approximate duration of the task in units of days (default)
- Task dependency (predecessors), for example, all tasks will have "Start of Project," Task 1, as a predecessor.
- Who will be doing that task and any other resources that may be needed, for example, 3D printers
- Some tasks may have a fixed start date because of capstone course requirements, for example, when the team's project presentations will be scheduled

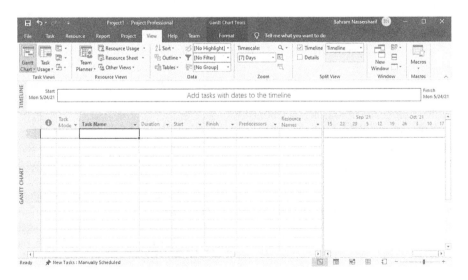

FIGURE 3.1 Project entry fields in MS Project.

Entering project information is straightforward. You start with a project starting date, which corresponds to when you have been assigned to a team to work on a particular project. Your capstone design course professor will assign you to a design team. An excellent way to enter project information is to designate every task as an auto-scheduled task. MS Project automatically calculates the task start and finish entries. All tasks should be made dependent on the "Start of Project" task by entering a "1" in the column with the heading "Predecessor." You enter all the tasks you have identified in the task column. For each task, include resources that are needed for those tasks, such as team members (TM1, TM2, TM3, TM4, etc.). You can use team member initials to save some space in your project plan charts.

Some tasks depend on previous tasks. For example, generating design concept solutions should define the problem, perform literature and patent searches, and create a plausible list of design specifications. Each task has a unique task number in MS Project to specify predecessor tasks by their number. As you estimate the duration of each task, you may need to make adjustments and iterate on your project schedule to ensure a Phase I end date for your project that matches the end of your academic term (semester or quarter). Your institution may have a one-, two-, or three-term capstone design requirement. So, the conclusion of Phase I may coincide with one or two terms.

Figure 3.2 shows an example project layout for a semester-long Phase I capstone course. You can adapt your own schedule from the example.

3.3.1 Independent and Dependent Tasks

Project types fall into two categories: dependent and independent. Independent tasks can be performed in parallel with one another. Dependent tasks, however, must be carried out in some sequence. Consider two tasks: Task 1 and Task 2. Let's assume that Tasks 1 and 2 are independent of each other. Figure 3.3 shows the independent representation of the two tasks. The two tasks can be performed in parallel without any interference. If two tasks are independent, they can be completed in any order with respect to each other: before, after, or in parallel.

Another situation for Tasks 1 and 2 is when they are dependent. Let us assume that results from Task 1 are necessary to perform Task 2. In this case, Task 2 cannot start until Task 1 has been completed. Therefore, Task 2 is dependent on Task 1 or Task 1 in the predecessor to Task 2. Figure 3.4 shows the dependent task diagram. Figure 3.5 shows a sequence of four dependent tasks.

Task networks can take any number of forms based on independent and dependent relationships. Figure 3.6 shows some of the configurations possible based on four tasks. Can you see how your capstone design project tasks could map into a network relationship?

The arrows that connect the task boxes may also have an attribute of time assigned to them to do more precise project planning. For example, we may want Task 2 to start after Task 1 plus some amount of time to allow for something to complete, such as shipping time for a part we purchase. Alternatively, the time associated with the connector arrow may be a negative number to indicate that we could start Task 2 a specified amount of time before Task 1 completes.

For example, a team member may become available if they finish their assignment early and start on the next task (Figure 3.7).

Project Management   45

		Task Mode	Task Name	Duration	Start	Finish	Predecessor	Resource Names
1		★	Star of Project	0 days	Mon 9/20/21	Mon 9/20/21		
2	👤	⇛	Organize and set up team meetings	1 day	Mon 9/20/21	Mon 9/20/21	1	TM1,TM2,TM3,TM4
3	○	⇛	▷ Weekly Progress Report	51 days	Fri 9/24/21	Fri 12/3/21		
15	👤	⇛	Schedule meeting with sponsor and meet	1 wk	Mon 9/20/21	Fri 9/24/21	1	TM1,TM2,TM3,TM4, Sponsor
16	👤	⇛	Review problem definition and prepare a list of questions	1 wk	Mon 9/20/21	Fri 9/24/21	1	TM1,TM2,TM3,TM4
17	👤	⇛	Search for additional information about the design problem	1 wk	Mon 9/20/21	Fri 9/24/21	1	TM1,TM2,TM3,TM4
18	👤	⇛	Literature search	1 wk	Mon 9/27/21	Fri 10/1/21	17	TM1,TM2,TM3,TM4
19	👤	⇛	Patent Search	1 wk	Mon 9/27/21	Fri 10/1/21	17	TM1,TM2,TM3,TM4
20		⇛	Create Design Specifications	1 wk	Mon 10/4/21	Fri 10/8/21	19	TM1,TM2,TM3,TM4
21		⇛	Generate design concepts	1 wk	Mon 10/11/21	Fri 10/15/21	20	TM1,TM2,TM3,TM4
22	👤	⇛	Analyze Concepts and Down Select	1 wk	Mon 10/18/21	Fri 10/22/21	21	TM1,TM2,TM3,TM4
23	👤	⇛	Prepare Presentation	1 wk	Mon 10/18/21	Fri 10/22/21	21	TM1,TM2,TM3,TM4
24		⇛	Critical Design Review	1 day	Mon 10/25/21	Mon 10/25/21	23	TM1,TM2,TM3,TM4
25		⇛	Select concept to model or prototype	3 days	Tue 10/26/21	Thu 10/28/21	24	TM1,TM2,TM3,TM4, Sponsor
26		⇛	Create prototype or model	4 wks	Fri 10/29/21	Thu 11/25/21	25	TM1,TM2,TM3,TM4
27		⇛	Document Design	2 wks	Fri 11/26/21	Thu 12/9/21	26	TM1,TM2,TM3,TM4
28		⇛	Final Presentation and Delivery	1 day	Fri 12/10/21	Fri 12/10/21	27	TM1,TM2,TM3,TM4
29		⇛	End of Phase I	1 day?	Mon 12/13/21	Mon 12/13/21	28	

FIGURE 3.2   Sample first project plan, including capstone design process steps.

3.4 ASSIGNING AND ACCEPTING RESPONSIBILITY

In a professional engineering organization, an organizational structure assigns the more experienced engineers as project managers. There is a manager that supervises a group of engineers on the project. The project manager has the responsibility to perform and complete the project successfully for the client. They also have the authority to allocate human and financial resources to perform and complete the project. The role of the project manager in the industry is not readily mapped into the capstone design projects.

In capstone design projects, a team of peers with equal responsibility and authority is working on the same project. This model requires that each member of the team be an active and productive participant in the project. The success or failure of the entire project falls on every team member's shoulders. Because of this unique model, team members volunteer to work on certain parts of the project, or the team

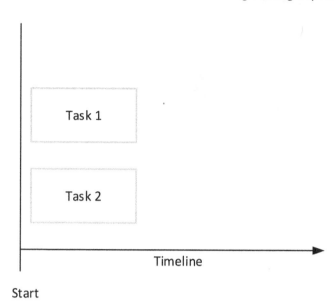

FIGURE 3.3 Illustration of independent tasks.

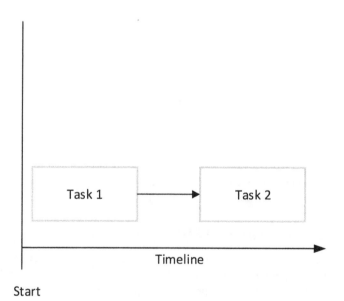

FIGURE 3.4 Illustration of dependent tasks.

may assign them the responsibility. Once the tasks have been divided and assigned to team members (self- or team-assigned), the team member must accept the responsibilities that come with that assignment. In this sense, every team member is a project manager on the team, supervising their role and contributions to the design project.

Project Management

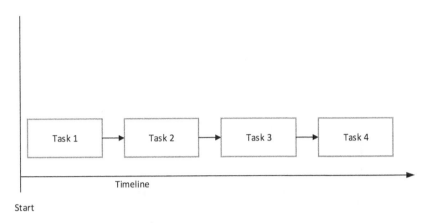

FIGURE 3.5 Illustration of multiple dependent tasks.

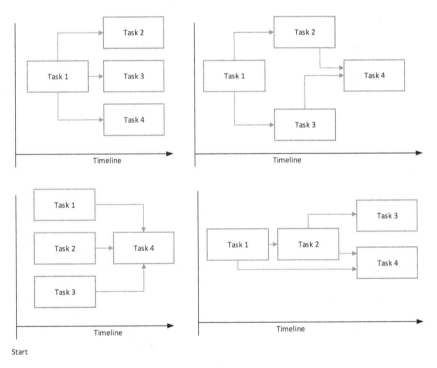

FIGURE 3.6 Illustration of combinations of independent and dependent tasks.

As a vested team member, you accept the responsibility to:

- Conduct yourself motivated by the project's goals and ensure the successful performance and completion of the design project. You must be willing to listen and compromise with team members, sponsors, and mentors.

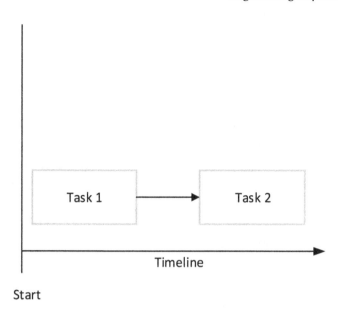

FIGURE 3.7 Illustration of dependent tasks.

- Understand and agree with the purpose of your project and work within realistic constraints, including schedule, resources, costs, and capabilities to realize your design project.
- Understand and reach a consensus with your team on the methods, tools, facilities, sequencing of tasks, scheduling, locations, division of tasks, technical approaches, etc.
- Work professionally, collaboratively, and with integrity with team members and other project stakeholders to reach reasonable compromises, mitigate conflict, and resolve problems throughout the capstone design project.
- Be motivated to achieve the best design and outcomes for the project.
- Have a positive attitude towards the project and fellow team members.

Attitude and motivation are the most important attributes of successful capstone design teams. You and only you can form and manage your attitude. If you have a positive attitude, it is reflected in everything you are doing on your design project. Your fellow team members, your sponsor, your mentors, and your professor notice and know when you exhibit a positive attitude. It is energizing for everyone around you. Similarly, it is noticed when you exhibit a poor attitude. If you have a poor attitude, it drains energy and enthusiasm from people around you.

Motivation is another important factor in successful projects. Motivation is derived from successes and feedback. Succeeding in project tasks and making progress are motivating. Some students are motivated by the prospect of achieving good grades or test scores. However, for capstone design, perhaps like other courses, good grades and scores are assigned to students who demonstrate excellent technique and rigor in their work and the results they achieved. Outstanding engineering design is not judged just as a result or product but also how you got there.

Some other tips for success in engineering design include:

- Be inspirational about your design project. Communicate to others why it is such a great project and how it will change the world.
- Look for and recognize obstacles on your path to success. Be proactive and remove those obstacles or find a workaround as early as possible.
- Enjoy your work on your design project and with your team members. There will be few opportunities in your professional life to work in such an environment.
- Exhibit good work habits. Always be polite, rational, and calm. Listen to your team members, mentors, sponsors, and professor and learn from what you hear. Respect the opinions and wisdom of others. Accept responsibility or blame when appropriate.
- Become an effective communicator interpersonally and professionally. Communicate important information during meetings and in written documents such as progress reports and interactions with the sponsors and mentors.
- Always act with complete integrity and ethics.
- Respect diversity and cultivate it to help your team achieve better results for your project.

3.5 WORKING WITH THE PROJECT PLAN

Working with a project plan is an iterative process. Once the tasks have been entered, the team will need to change and update the plan to add details and progress information. The project plan can also assist the team in financial analysis and cost estimation for the project. The project views in MS Project can be utilized to analyze the entire project and identify where additional information will need to be added.

You may need to change the start date of the plan rather than using the current date. When you create a new project in MS Project, the start date for the project is set to the current date. To change the start date, click on the Project tab, and then click on Project Information in the properties group, as shown in Figure 3.8.

While you are setting the date, you can also choose what type of work calendar is appropriate for the project and team members. You will need to create a customized calendar appropriate for your team members' schedules. As a student, your availability to work on the design project is different than a typical professional work calendar. Each team member can enter their personal work calendar into MS Project. Use the "Change Working Time" option in the Project ribbon tab to access the Calendar and Project Options. Each team member can indicate exceptions to the default calendar, for example, traveling to a student conference. Figure 3.9 shows the Calendar options. Project options can be accessed from the same window, as shown in Figure 3.10.

3.5.1 Resources

Project resources consist of people, equipment, and facilities. People affiliated with a project include the student members of the team, mentors, sponsors, and professors. Each person resource should be set up with appropriate data regarding availability to work on the project and cost (pay rate).

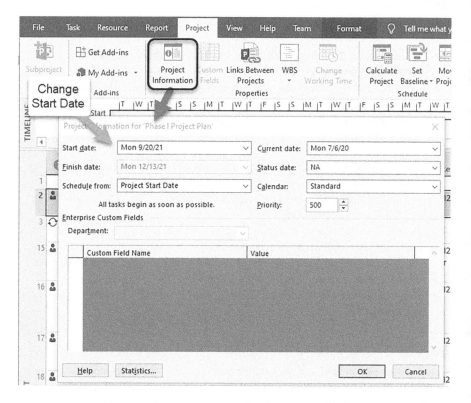

FIGURE 3.8 Changing project start date in Project Information.

In the university environment of capstone design, people do not usually get paid to work on the design project. Still, the information for the pay rate is helpful in the financial analysis and value of the design project. Laboratory, machine shop, computers, and software on the project also have an inherent value to the project, so they should be counted as part of the project costs and financial analysis. Supplies and materials used in the project are also counted as project resources and contribute to the financial analysis for the project. MS Project allows the inclusion of any resource in the project, which helps significantly create a calculated financial analysis for the project.

Resources are set up in MS Project from the "Resource" tab, as shown in Figure 3.11.

3.5.2 Gantt Chart

A common way to view and share project information is by using the Gantt chart format, which is also the default in MS Project. A Gantt chart shows the project schedule and the dependencies of the tasks, resources, and milestones in a graphical bar chart form. The chart was designed by and is named after Henry Gantt (sometime around 1910–1915).

A sample Gantt chart for Phase I of the capstone design is shown in Figure 3.12. The time scale on the chart can be modified by double-clicking in the timeline bar above the

Project Management

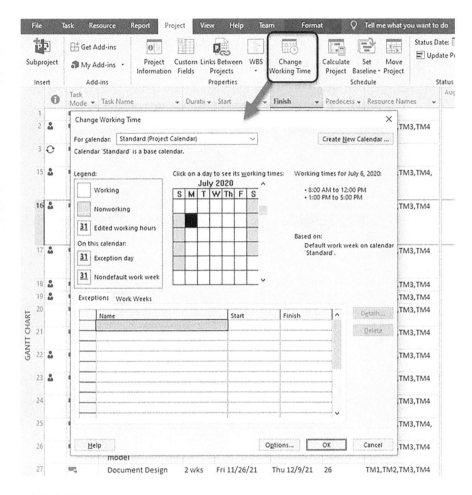

FIGURE 3.9 Changing working time for a resource.

chart. For working meetings and communications with the sponsor, a day resolution for the chart is probably preferred. We have chosen a monthly resolution to fit the chart for publication purposes of this book. Gantt charts are frequently printed on a large format paper to show the details of the project. Such charts are printed and mounted on a conference room wall in the meeting rooms of engineering companies. Capstone project teams should also print their Gantt charts on the largest format printers available.

The Gantt chart should be formatted with color to enhance the visual communication of information.

3.5.3 Network Diagram and Optimizing the Project Plan

The task dependencies in a project plan form linkages between tasks. This was evident in the sample Phase project plan in Figure 3.12 in the form of arrows connecting the taskbars. The dependency information becomes more evident and beneficial

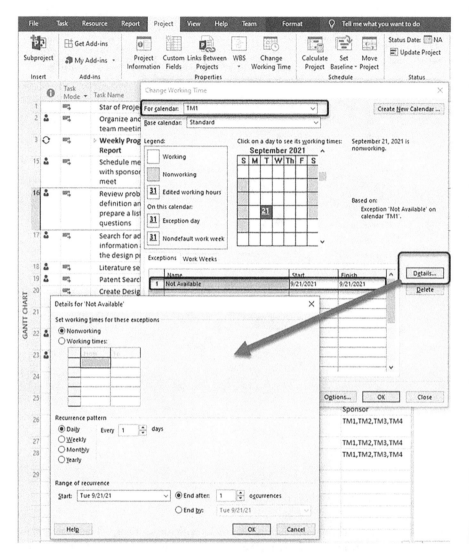

FIGURE 3.10 Team member customized calendar.

by using the network diagram feature in MS Project. Figure 3.13 shows the network diagram for our sample Phase I project plan.

The chain of gray task boxes in the figure shows a critical path in the project. The task path analysis can also be done in the Gantt chart view by formatting the taskbar styles to the task path format. The critical path in a project represents the minimum amount of time in which the project can be completed. The critical path analysis should be done throughout the project as the project plan is updated to optimize the project execution. Tasks that fall on the critical path should be examined

Project Management

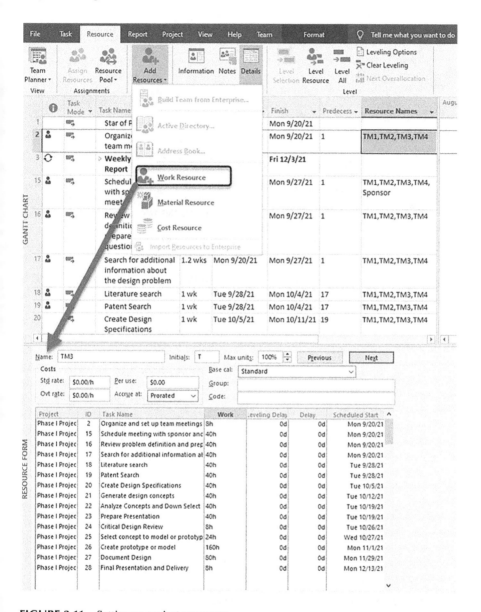

FIGURE 3.11 Setting up project resources.

to determine if they can be shortened or decomposed into activities that can be performed in parallel, perhaps by multiple team members.

It may be possible to start a task on the critical path earlier than when it is planned. The task dependencies can be fine-tuned by examining opportunities for parallel task execution. Figure 3.14 shows the critical path on the Gantt chart by using the Format tab and checking the Critical tasks box and the Slack box. Double-clicking on a link between the two tasks allows us to change the lag time (positive or negative).

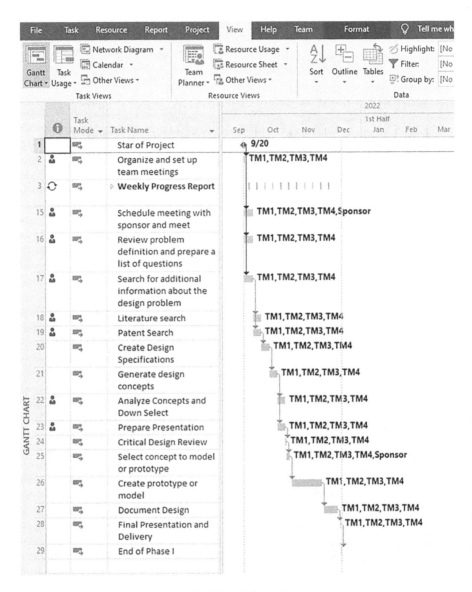

FIGURE 3.12 Sample capstone design Phase I Gantt chart.

We can also change the task dependency. In Figure 3.14, we have changed the lag to −2, which allows the start of the creative design specifications task to overlap with the patent search for 2 days.

The slack analysis is indicated by the underlining of the resources in the figure. A task with slack can be started later than planned or finished later than planned without affecting the project's total duration. The design teams must perform the critical path analysis to ensure the completion of the project before the end of the academic term.

Project Management

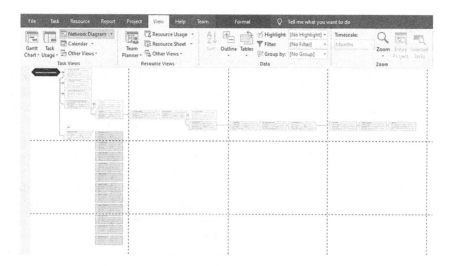

FIGURE 3.13 Sample capstone design Phase I network diagram chart.

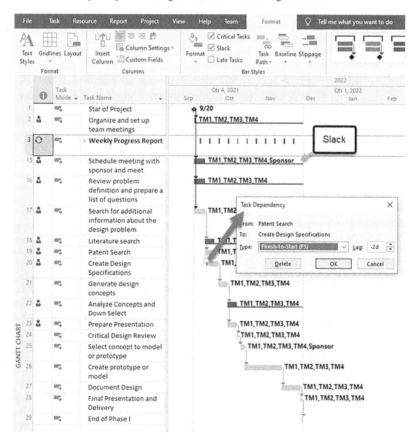

FIGURE 3.14 Critical path and tasks with slack on the Gantt chart.

3.6 RESOURCE LEVELING

According to the calendar for that resource, when you assign resources to a task in MS Project, the default commitment is a full-time commitment to that task. For example, if you assign team member 1 (TM1) to the assignment for patent search, then the default assumption is that TM1 will spend 8 hours a day on that task. The default assumption for time commitment is probably not valid. So, you will have to adjust the number of hours or percent of the time that TM1 will work on the task. Simultaneously, TM1 can be assigned to multiple parallel tasks. The actual time commitment of TM1 will have to be adjusted in the project plan such that they are not working more than their total available time on any given day. If you remove a responsibility from TM1 to lighten their load, then that task will have to be assigned to another team member. In this manner, the workload must be leveled across all available resources allocated to the task. This iterative process for balancing and equalizing the workload is called load leveling or resource leveling.

Initial data for team member participation may be estimated or left as a default. Still, they must be corrected when the team reviews the project plan carefully and starts working with the project plan. Multiple views of the project plan are available to simplify the task of fine-tuning the project plan. Figure 3.15 shows the task usage view in MS Project. Because the default calendar in MS Project is set up for a 40-hour workweek, all task data entered defaults to the 40-hour per week. For students in capstone design, a reasonable estimate for the workload is 12 hours per week. Some weeks may be lighter, and others may be higher depending on how well the team distributes and assigns their tasks. The task usage view is where corrections to workload estimates for team members can be viewed and corrected.

Another valuable view of the project is the team planner view (see Figure 3.16). In this view, the team members can see their task assignments on the project timeline. They can compare their assigned schedules to their other commitments (such as other classes or work) and adjust their engagement with the project tasks correctly.

In the resource usage view, team members can check and see if they have been overallocated for their time daily. Overcommitments should be resolved as quickly as possible to avoid bottlenecks and unnecessary disappointment in the project's progress. Figure 3.17 shows where to look for daily commitments.

The resource sheet view (see Figure 3.18) can be used to enter the hourly rate for individuals affiliated with the project. This information will be used later by the team to perform a financial analysis of their design project. In addition to the cost for individuals, the cost for any other resources such as computer software, materials, machines, travel, shipping, and services can be included in the resource sheet and on the project plan to help with the accounting of resources and financial analysis of the project. Engineering projects typically are required to include such detailed accounting in their regular reporting to their clients or sponsors.

Another useful view of the project is the calendar view, which is shown in Figure 3.19. The team can use the calendar view as a reminder for tasks that currently need to be performed as well as tasks coming up during the next weeks. It is a useful view for progress tracking and reporting.

Project Management

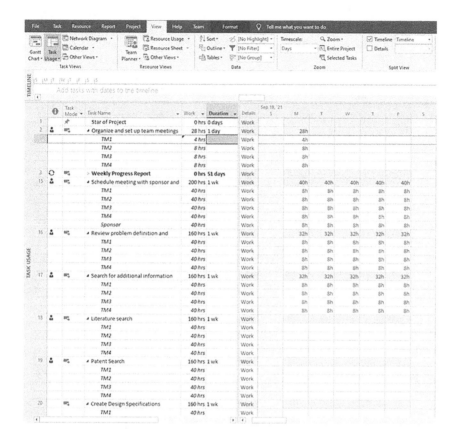

FIGURE 3.15 Task usage view.

3.7 PROJECT REPORTS

A good project plan should guide the entire execution of the project. The initial project plan created and refined (also called a base project plan) by the team will track the progress on each task. A reasonable interval for updating task level progress is weekly. The project plan should be updated before the team submits its weekly progress report and attached.

Actual project performance may be different than the base plan. For design project plans, some guiding questions can help the team members analyze their plan:

- Is the plan on track? Will the project be done before the deadline? Has the duration and cost changed? What has changed?
- Have there been any changes in resource availability (people, funding, equipment, software)?
- Are the capstone course milestones and requirements achievable with the current project plan?
- What changes are needed to the project plan? Can the plan be further optimized?

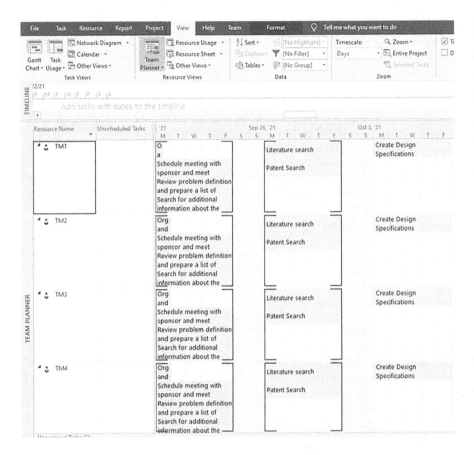

FIGURE 3.16 Team planner view.

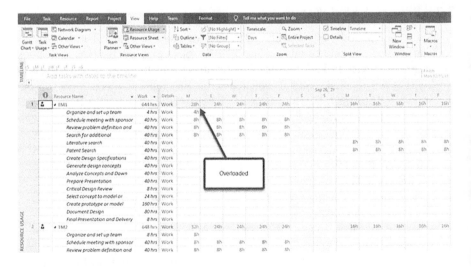

FIGURE 3.17 Resource usage view.

Project Management

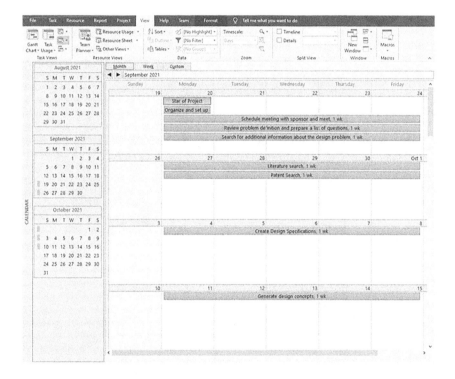

FIGURE 3.18 Resource sheet view.

FIGURE 3.19 Calendar view.

Answering the above questions will help the team measure progress and identify changes or corrections that are needed. The report tab (see Figure 3.20) provides several report view of the project that is incredibly informative and useful for progress reports, design presentations, and design reports.

Cost is a critical driver in the course of design projects. The cost view report provides a quick summary of the project costs. Figure 3.21 shows the cost overview report. A companion report to the cost overview is the cash flow report (see Figure 3.22). The summary results in the project reports are updated as soon as any changes are made to the information entered for tasks and resources.

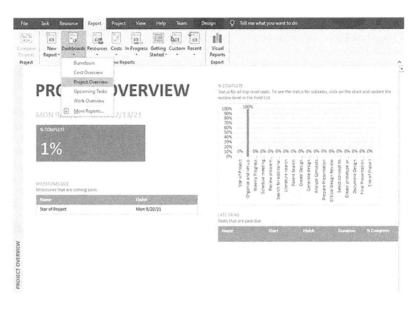

FIGURE 3.20 Project overview report.

FIGURE 3.21 Project cost overview report.

Project Management

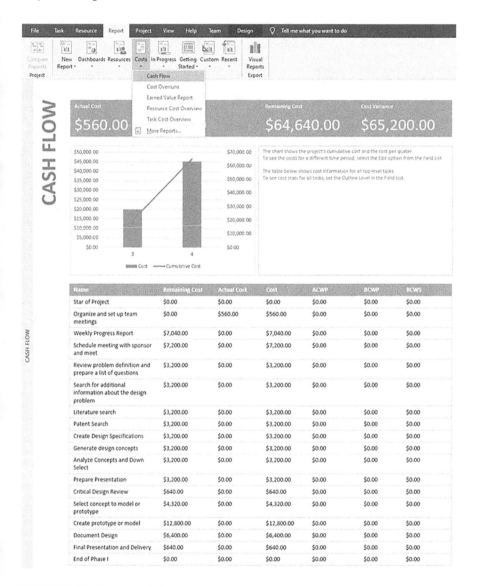

FIGURE 3.22 Project cash flow report.

4 Defining the Design Problem

Capstone design problem ideas come from a variety of sources. In all cases, there is a sponsor or a motivator for the project. The motivation for the problem comes from a need or desire to create/design something that will solve a problem, fix a perceived defect or deficiency in a system or device, or make something better. It starts with a vague notion of what needs to be done. The vagueness is desired for capstone design problems because it represents an open-ended problem where many solutions may exist. The best solution for the given (or not given) constraints is desired. Because the initial problem statements are incomplete, the first task for the design team is to learn and understand the problem domain and resolve the unknown or unresolved aspects of the problem through critical thinking, search, research, and iteration.

Often, the sponsor does not know the answers to the team's questions at the start of the project. Some questions cannot be answered initially because the design team is not fully knowledgeable in the particular problem domain, methods, or technologies involved.

Defining the problem is probably the most critical aspect of any project. Problem definition is new to most engineering students because they have been presented mostly with entirely defined problems in their courses before capstone design. There is usually only one correct solution to the problem. It is difficult to define a problem if you do not know what is missing from the initial problem definition or expression of some need. Many of the missing pieces will be identified as the team engages in problem-solving.

In most cases, the team will need additional information to understand their design problem better and identify missing pieces. Additional information may be available from the sponsor. An interview with the problem sponsor will reveal other facts and information about the problem and its motivation. However, the sponsor may not have all of the needed information, and usually, they do not have it.

Other sources of information include interviews with experts, a survey of potential users, a search of the patent databases for related ideas, a search for relevant articles and papers, an internet search for similar problems or solutions, investigative search, and brainstorming.

4.1 METHODS FOR UNDERSTANDING OPEN-ENDED PROBLEMS

There are many approaches to trying to understand open-ended problems. We will explore some of those techniques here. The approaches discussed here will be a reasonable starting point for collecting more information about the design problem.

4.1.1 Brainstorming

Brainstorming is probably the oldest and best-known method for group problem-solving. Alex Osborne introduced the term brainstorming in his 1953 book entitled "Applied imagination; principles and procedures of creative thinking." He described how an organized group of people could generate more and better ideas than if those individuals were working on their own. He explained, "brainstorm means using the brain to storm a creative problem and to do so in commando fashion, each 'stormer' audaciously attacking the same objective."

Creative thinking is an essential part of capstone design. The design team is expected to apply creative thinking methods throughout their design project. Innovative thinking by the team members is also part of team building and working together.

- Osborne introduced some powerful ideas for a brainstorming activity.
- Focus on the number of ideas rather than the quality of each. Disallow criticism of ideas shared.
- Encourage far-fetched ideas.
- Expand and build on ideas presented by fellow team members.

Generating many ideas (including assumptions and questions) at the start is very useful. Producing a vast number of concepts and ideas pushes the team to think outside the box. One of those ideas may be the design project that your team will be working on for the rest of the capstone experience.

Withholding criticism is an enabler of creative and outside-the-box thinking. Criticism tends to shut out many good ideas because team members will be more reluctant to share their thoughts if they are being ridiculed.

Encouraging far-fetched ideas is an enabler for creative thinking. It promotes innovation and thinking outside the box. Perhaps many of those ideas will be eliminated by the team later, but there is a potential for a significant breakthrough or an entirely new concept.

Let's look at an example. Last year, one of my design teams was tasked to undertake a project to design a five-axis 3D printer. The design problem was proposed by a sponsor at the Naval Undersea Warfare Center in Newport, Rhode Island. The motivation was to design a better 3D printer capable of printing some geometries without the dependence on a fixed z-direction.

Example 4.1 – Team of five students assigned to
Attributes of a perfect problem definition for a design project include:

- The design project will satisfy every need and solves every problem.
- The design can be achieved quickly and manufactured by the team.
- The design will not require maintenance, calibration, and upkeep.
- The design will require no training (or minimal training).
- The user will not be required to change behavior to use the design.
- The design will save materials and money for the sponsor.
- The design will be mistake-proof.

Defining the Design Problem 65

On the other hand, poor problem definition will result in:

- Over-design by wasting materials and resources.
- Too much precision required (unnecessary tight tolerances).
- Untested solution methods outside of the expertise of the team members.
- Too much or unnecessary functionality. Too limited functionality.
- Requires the user to change behavior or put too much effort into using the design.
- Too expensive.
- Limited use or availability design.

Scrutinize your problem definition to stay away from trouble areas and make it easier and more efficient for the entire team to achieve a successful design.

4.2 GATHERING INFORMATION

The team must know and understand the design problem before an applicable definition is achieved such that the problem can be solved. One way to better understand the design problem is to look for the motivation or why the design problem was posed. Understanding the motivation (user requirements) is essential in developing a set of comprehensive design specifications. The customer, sponsor, or user of the design will be a good source of information. Several methods can be used to gather information about the desired features in the design.

Customer/user/sponsor interviews. As soon as the team is formed and a meeting is arranged with the sponsor, the team should develop plans to interview the project's stakeholders. Typically, the first activity arranged by the sponsor is to visit their company, lab, etc. During that first meeting, the team should ask potential users or customers about their design. Will the design be used internally, or will it become a product that the company will produce/manufacture to provide/sell to their customers? The team should develop a list of questions for the interview, such as: What benefit will the new design bring to their work environment? What is deficient or could be done better? What features will be of interest? What are the safety considerations to be included in the design?

Group interviews. The interview with users/customers can be done as a group. There is some synergy in a group meeting where people build on each other's ideas and concerns. A team member can be the moderator to keep injecting new questions when the discussion stalls or needs to get back on track. It is essential to take notes during the interviews. With the permission of individuals attending the meeting, you may be able to record the session and refer to it later. A video recording may be more informative than just a voice recording.

User/operator/customer surveys. Surveys are an excellent tool to collect information from a large number of people. Several tools are available to create and deliver the survey, e.g., SurveyMonkey and Qualtrics. Survey questions should be constructed carefully to avoid bias or multi-barrel questions where it is unclear how to interpret the answer to that question. The list of recipients for the survey must also be assembled carefully to ensure the validity of the survey results.

4.2.1 Surveys

Suppose the design project is something many people will use (e.g., a software application or a product). In that case, a survey can help the team gather information about customer/user needs and desires. Sometimes survey research is needed to collect data from experts in a field.

Think about what you need to learn from the survey, i.e., what is the purpose of the survey? What information can you learn from the stakeholders/users/customers for your design specifications or requirements?

Google Forms is an excellent tool for creating and administering a simple survey. Google Forms is free and readily available at universities as part of their G-Suite implementation. It may also be appropriate to perform the survey as an interview (phone, video-conference, or in-person). If only a handful of people will be surveyed, then an interview format may work best.

Figure 4.1 shows the process for surveying customers, stakeholders, operators, or potential users. Designing the survey is a process of group thinking, and the sponsors of the project can also help identify relevant questions and variables. The first step is to make a list of variables that need to be measured in the survey. Consider the following example:

Example 4.2 – An exciting design project is to design an automated pill dispenser. Many individuals, particularly the elderly, have several medications in pills, some prescription and some over-the-counter. Keeping track of the frequency of each tablet becomes a challenging task for many patients. Most people have difficulty remembering if a medication was taken or what time to take what medication. Some method of reminding is needed. The design challenge is to create a device that will be programmable to dispense pills, send reminders to the patient, and keep a log of medication taken.

A survey is beneficial in determining the design specifications for such an automated pill dispenser. The team will need to decide who will need to be surveyed: some sampling of patients, medical doctors, pharmacists, and caretakers. The team

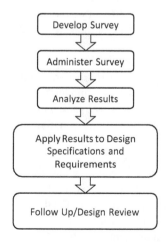

FIGURE 4.1 Survey process flow.

Defining the Design Problem

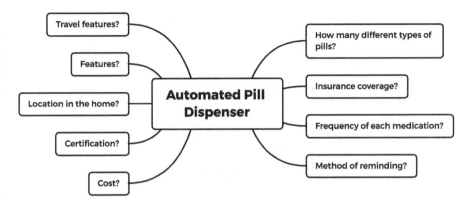

FIGURE 4.2 Automated pill dispenser survey variables/questions relationship diagram.

will also need to identify what variables will be included in the survey. Figure 4.2 shows a relationship diagram for developing survey questions or variables.

Once the team has developed a list of variables/questions, a survey method can be selected based on the number of expected participants. Regardless of the method, the same set of questions should be asked of each participant. You can also include open-ended questions or ask for comments from the participants. The survey should start with a preamble to explain the motivation behind the survey and how it could benefit the design project and, therefore, the survey participants.

Before you send out your survey to the actual participants, create a trial version and have your team members, friends, family, fellow students, sponsors, and mentors review it for you and give you constructive comments to improve your survey instrument. After making corrections from this initial review, it may be necessary to do a second trial to further streamline your survey.

Figure 4.3 shows the steps involved in creating your test survey. If you are surveying more than ten people, an electronic survey may work best to collect the

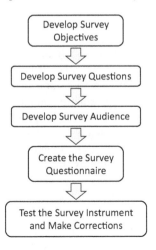

FIGURE 4.3 Develop survey.

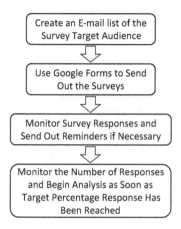

FIGURE 4.4 Administer survey.

information you need quickly and efficiently. Any automated survey tools can be used, such as SurveyMonkey, Qualtrics, or Google Forms. Google Forms is free and is the simplest way to conduct a survey. SurveyMonkey and Qualtrics are much more powerful tools, but they also have a learning curve to utilize their feature set for post-survey analysis fully.

Figure 4.4 shows the steps involved in administering your survey. You will need a list of e-mails for people whom you will ask to participate in your survey. You may need to select your survey participants. You include diversity in age, gender, race, and other demographics to sample the responses you need to create your design adequately.

Once you complete administering your survey, you analyze the responses statistically and create a report included in your final design report and share it with your sponsor. The survey results should help your team develop a more detailed set of design specifications to guide your design efforts.

4.3 SOURCES OF INFORMATION

Information needed to develop further the definition of the design problem and the necessary design specifications can be gathered from multiple sources. Figure 4.5 shows the significant sources of information for capstone design projects.

Sponsors may have additional information about the design project, motivation, and details that were not included in the initial problem definition provided as the starting point. It will be the design team's responsibility to interact with the sponsor to discover any additional information. Sometimes the main point of contact for the sponsor (for example, a company providing the design problem) may not be the expert with the design problem. Perhaps another member of the sponsor company may have and be able to provide additional details.

Domain experts could be another source of information for the design project. Domain experts are technical professionals (including members of the faculty at a university) who may have done research or have work experience in the field of the design project. They may provide additional insights or documents (research articles,

Defining the Design Problem

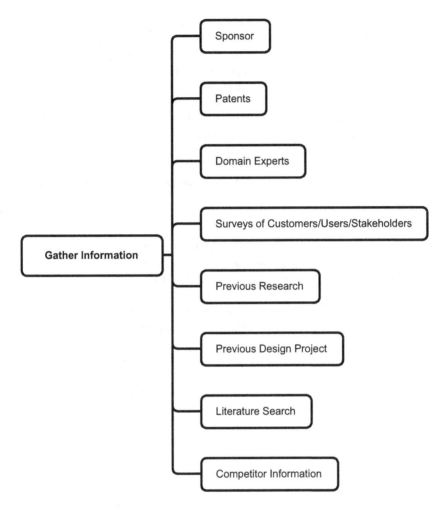

FIGURE 4.5 Sources of information for design projects.

books, journals, reports, etc.). The design team should network through their professor, mentors, and sponsors to find the domain experts.

The design project may be a derivative of some previous research and development work. The relationship to previous research may have been stated as part of the problem definition, or it may have to be discovered by the design team. Previous research may have been documented in a journal article, conference paper, report, or part of a book. The design team will be responsible for finding, reading, and understanding any previous research work related to the design topic.

Sometimes design projects are based on previous design projects. They could be a derivative of design work performed previously or a continuation of a past design project. The previous design work, especially if it were another capstone design project, would have been documented in a design report. The design team should locate, read, and use the previous design report before launching into their design project.

Many design projects are derived from a desire to design a better product or process than existing and is in use. There may be many similar (competitor) products. The design team should research and find as much information as possible on the competitors. Companies research their competitors all the time. They purchase or acquire competitor products or processes and dissect them to understand what makes them useful or how they work. Dissecting competitor products or processes is a practical technique for the design team to learn from what has already been accomplished. Competitor information can usually be found by performing an internet search. It may be necessary to acquire the product or software to understand better the design features and techniques used.

We will discuss the patent information and literature search in the following sections.

4.3.1 Literature Search

A literature search is a required step in any scientific or engineering research and development project. It is also a required element of a design project that relies on past research, discovery, studies, or design projects. Conducting a literature search means collecting and compiling information in the form of books, articles, papers, presentations, reports, theses, or dissertations related to the topic of your design project.

Collecting, reading, comprehending, and integrating information from a literature search provides a background, methodologies, comparisons, context, and identification of knowledge gaps that will assist the design team in performing their design work with credibility and rigor.

The process flow for a literature search is shown in Figure 4.6. The process begins with brainstorming topics and keywords related to your project. If you have a report or article provided to you, it will be an excellent starting point. Look through the report or article you were provided and develop your list of topics from that source. References to the report or article are an excellent source for topics and keywords as well. Your sponsor may also be able to assist in providing topics and keywords for your literature search.

The next step is to find sources of literature that you will search. Your university libraries will have information and tutorials for you to perform a literature search of the databases that they have purchased and support. They will provide the ability to download some articles. You can also check out books that the library may have or interlibrary loan agreements with nearby or affiliated universities.

In addition to the library services through your campus, you can perform an online search of some free sources, including scholar.google.com, academia.edu, and www.researchgate.net.

You may have to create an account to use online services. Some provide additional features for fees. The free search should be sufficient for what you need for your design projects.

Searching through the databases will produce many or few results based on the keywords and breadth of topics that you included in your search. You can usually tell from the title or abstract of each article if it is relevant for your design project.

When you find an article or book relevant to your project, you can expand by reading the article or sections of the book and looking through their reference list

Defining the Design Problem

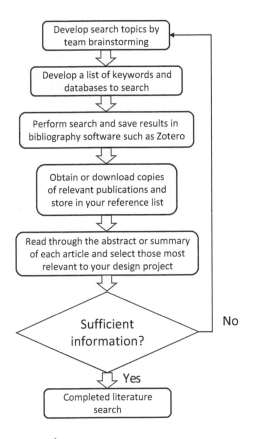

FIGURE 4.6 Literature search process.

to find additional related articles. You should identify and gain access to as many articles and books as sources of information for your design project. The best sources are publications in peer-reviewed journals and books. The peer reviews conducted to publish scientific and engineering journal articles and books support the validity of the information provided. Be careful about using non-peer-reviewed internet documents or blogs. There is much misinformation woven into the fabric of the internet.

The ability to conduct a successful literature search is a skill that will serve you well for your future professional career or graduate education. So, spend the necessary time to do this well.

4.3.1.1 Literature Search Assignment

See the electronic file assignment for conducting a literature search for your design project. The assignment file guides you through performing the literature search for your design problem. The information you will collect as part of that assignment will be helpful for conducting the technical work on your project and completing future steps in your design activities.

4.3.2 Patents and Patent Search

What is a patent? The definition for a patent is provided by the United States Trade and Patent Office (USPTO) website as: "A patent for an invention is the grant of a property right to the inventor, issued by the United States Patent and Trademark Office. Generally, the term of a new patent is 20 years from the date on which the application for the patent was filed in the United States or, in special cases, from the date an earlier related application was filed, subject to the payment of maintenance fees. U.S. patent grants are effective only within the United States, U.S. territories, and U.S. possessions. Under certain circumstances, patent term extensions or adjustments may be available.

The right conferred by the patent grant is, in the language of the statute and of the grant itself, "the right to exclude others from making, using, offering for sale, or selling" the invention in the United States or "importing" the invention into the United States. What is granted is not the right to make, use, offer for sale, sell, or import, but to exclude others from making, using, offering for sale, selling, or importing the invention. Once a patent is issued, the patentee must enforce the patent without aid of the USPTO."

So, a patent is a legal right granted to the inventor to exclude others from using their invention unless the inventor reaches a legal agreement with interested parties to allow (exclusively or not) them to create products or services commercially. The inventor may sell or reach a licensing deal with interested parties and profit from their invention.

Many capstone design projects have the potential to generate new inventions. If the project sponsor is a corporate entity or an individual entrepreneur, then they will be interested in having full rights to the intellectual property generated from the design project. While it is possible to engage in negotiating intellectual property rights with corporations, it is a deterrent for some companies to work with universities. Regardless, any agreement regarding the sharing of intellectual property rights must be negotiated at the start of the design project.

U.S. laws provide for three types of patents: utility, design, and plant patents. Utility patents are granted to inventors of any "new and useful process, the machine, article of manufacture, or composition of matter, or any new and useful improvement thereof." Design patents are granted to inventors of "new, original, and ornamental design for an article of manufacture." Plant patents are granted to inventors of "a distinct and new variety of plant reproduced asexually."

In capstone design, we are interested in utility and design patents because of their relevance. Past inventions may provide clues and information in the design project. Figure 4.7 shows the number of U.S. patents issued per year since the start of the patent office in 1790. There has been a significant increase in the number of utility patents issued over the past decade. The new ideas present in the patent database is a fantastic resource for designers both to learn from what has been discovered and to check if your ideas are unique and not patented previously.

Searching the patent database is different than conducting an internet search, so it requires knowing how the patent database is structured and how to properly explore the database to obtain useful results for your project. Some of the patent databases have been captured by Google and is available through scholar.google.com. It can be searched using the same methods that you know how to use.

Defining the Design Problem

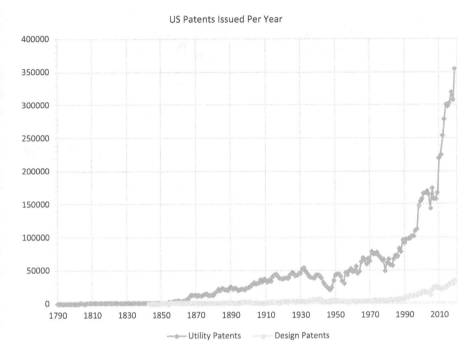

FIGURE 4.7 The number of US Utility and Design patents issued per year since 1790. (U.S. Patent and Trademark Office.)

The U.S. patent office recommends using a seven-step strategy for searching the patent databases.

Step 1 is to brainstorm keywords that best describe what you are interested in based on the purpose, composition, and use. This step is challenging to many students because the first inclination is to think of keywords similar to what you would use for a Google search. This bias in thinking of keywords will produce poor results with a patent office search. Focus on keywords that describe the purpose, composition, and use of your design. For example, consider the automated pill dispenser that we introduced in Example 4.2. After some brainstorming, the students on that team came up with the following keywords: device or container for dispensing pills and tablets.

Step 2 is to use the keywords or description of the use or function that you listed under step 1 and perform a classification search at www.uspto.gov. In the site search box, type in "CPC Scheme," and then the keywords or description. CPC stands for Cooperative Patent Classification, which is based on the International Patent Classification (IPC). For example, for the pill dispenser, type in "CPC Scheme device or container for dispensing pills and tablets." Figure 4.8 shows the USPTO home page and the search box for entering the keywords or descriptions. Figure 4.9 shows the search results. If the search results are not helpful, consider changing your keywords or description. Keep in mind that this method of search is different than the typical Google search. If the results are not helpful, another option is to search the world intellectual property organization (https://www.wipo.int/classifications/ipc/ipcpub/?notion=catchword) site for catchwords.

FIGURE 4.8 The first step in patent search, classification search.

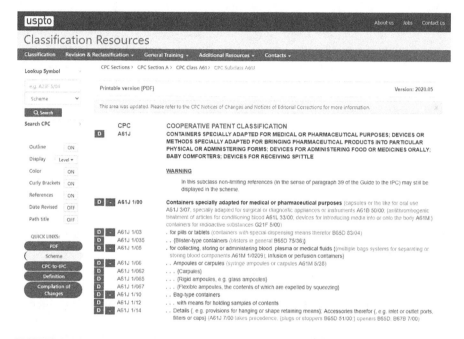

FIGURE 4.9 Results from classification search for pill dispenser.

Step 3 is to find some applicable CPC or IPC classification numbers to advance your search further. You will need to be persistent about looking through the results returned to find valuable results. After all, this is not a Google-style internet search! In the case of our search for the pill dispenser-related patents, we found class A61J: CONTAINERS SPECIALLY ADAPTED FOR MEDICAL OR PHARMACEUTICAL PURPOSES; DEVICES OR METHODS SPECIALLY ADAPTED FOR BRINGING PHARMACEUTICAL PRODUCTS INTO PARTICULAR PHYSICAL OR ADMINISTERING FORMS; DEVICES FOR ADMINISTERING FOOD OR MEDICINES ORALLY; BABY COMFORTERS; DEVICES FOR RECEIVING SPITTLE.

Step 4 is to search the Application Full-Text and Image Database (AppFT) (http://appft.uspto.gov) using the CPC classification number. Figure 4.10 shows the Patent Full-Text Databases home page. Links are provided to both patents issued and patent applications. Full-text search is available from 1976 when the USPTO changed the way that they store the patent information. Figure 4.11 shows the flow chart for a patent

Defining the Design Problem 75

FIGURE 4.10 USPTO Patent Full-Text Database.

search which can be used to navigate the process. The most valuable links are "Quick Search" and "Advanced Search." A quick search provides for a simplified search of terms with simple Boolean operations: "AND," "OR," "ANDNOT." Figure 4.12 shows the quick search screen.

The next step is finding relevant patents related to your project from the quick search result list. Finding relevant patents is not guaranteed, but usually, ideas can be stimulated by reviewing what has already been invented (for example, see Figure 4.13). If you do not see any relevant patents, then you should re-examine your keyword list. Should you eliminate some keywords based on the results you see? Do you have enough keywords? Should you modify some keywords?

You can refine your search by using "Advanced Search" if you have too many results or have learned enough information from "Quick Search."

4.3.2.1 Patent Search Assignment

Consider your design project. Follow the steps described in this section to create a list of relevant keywords to conduct a patent search. See the electronic patent search assignment document provided for guidance on the steps to conduct your search and document the results for your project. You should coordinate with your team to agree on a standard set of terms and then divide the patent search task so that each team member can experience the process and develop a set of results that the team will share.

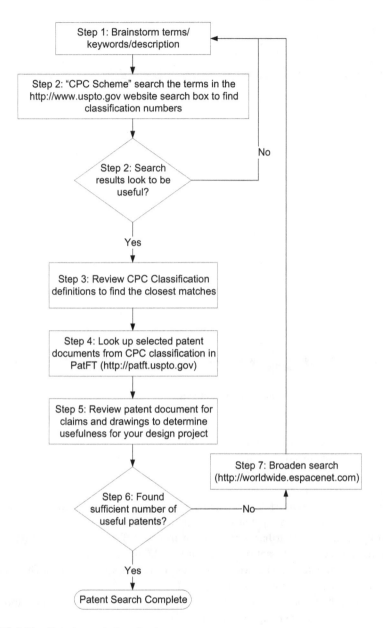

FIGURE 4.11 Patent search flowchart.

4.4 DESIGN FOR X

Design is an iterative process. The first idea or rendition of a design solution is hardly ever the final design. Quality in design is one of the attributes that require iteration. In any iteration, we need to establish some objectives to test against those objectives to check for convergence or knowing when to stop the iteration. It is best to develop

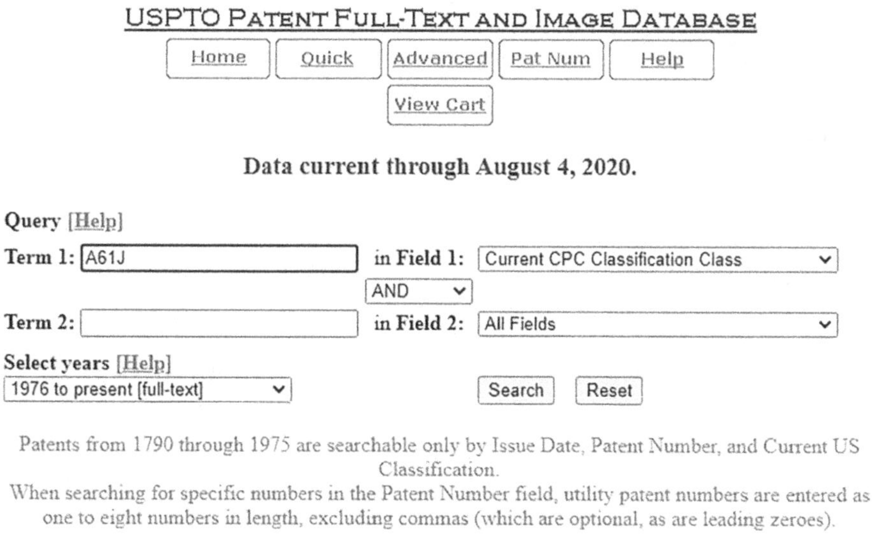

Defining the Design Problem 77

```
         USPTO PATENT FULL-TEXT AND IMAGE DATABASE
              [ Home ]  [ Quick ]  [ Advanced ]  [ Pat Num ]  [ Help ]
                              [ View Cart ]
```

Data current through August 4, 2020.

Query [Help]
Term 1: [A61J]   in **Field 1:** [Current CPC Classification Class ▾]
                            [AND ▾]
Term 2: [   ]   in **Field 2:** [All Fields ▾]
Select years [Help]
[1976 to present [full-text] ▾]   [Search] [Reset]

Patents from 1790 through 1975 are searchable only by Issue Date, Patent Number, and Current US Classification.
When searching for specific numbers in the Patent Number field, utility patent numbers are entered as one to eight numbers in length, excluding commas (which are optional, as are leading zeroes).

FIGURE 4.12 Classification search using Quick Search.

important objective attributes or functions before starting the design solution activity and integrating those into the process.

Design for X is a methodology to apply and use the knowledge gained by designers in various fields in your design activity. The X in Design for X is a variable that can take on many different values such as cost, manufacture, assembly, maintainability, operability, quality, reliability, safety, sustainability, ergonomics, installability, user-friendliness, portability (software), modularity, etc. Consideration of a set of X values during the design process will lengthen the time to achieve a design solution but has the considerable benefit of resulting in a better design.

You will have to decide which X's apply to your design and incorporate those into your design process. For a typical capstone design project, you should limit the number of X values to three or four, which are most important for your design objectives. It is also desirable to pick X values to demonstrate how you incorporated those methods into your design activity. You have to show, explain, or demonstrate how your design achieves the X value.

For example, suppose you selected "manufacture" as one of your X values. In that case, you have to explain and demonstrate how your design was created (or changed during redesign) to achieve manufacturability. In many cases, the design team is responsible for creating the prototype of the design. Did you design your device or system so that you could create the prototype design? If it is a physical device, will you make your design and show that it works? If it is a software program, will you show that you wrote the software, and because of your design methods, it made it easier to develop the software? If it is a process, can you explain how you designed the process so that you could implement it? Were you able to reduce the number of steps in creating your design? Were you able to reduce the number of parts or modules? Were you able to use off-the-shelf parts, modules, or algorithms? Were you

Results of Search in US Patent Collection db for:
CPCL/A61J: 11392 patents.
Hits 1 through 50 out of 11392

[Next 50 Hits]

[Jump To] [   ]

[Refine Search] [CPCL/A61J   ]

PAT. NO.		Title
1	RE48,136	Pill dispensing method and apparatus
2	10,733,397	Method for managing at least one container and associated methods and devices
3	10,732,083	Thawing biological substances
4	10,730,687	Intelligent medicine dispenser
5	10,730,682	Connecting and container system
6	10,730,207	Process for manufacturing a resulting pharmaceutical film
7	10,730,058	Apparatus, systems and methods for storing, treating and/or processing blood and blood components
8	10,729,859	Mask for administration of inhaled medication
9	10,729,854	Housing for mounting a container on an injection pen, assembly forming an injectable product reservoir for an injection pen and injection pen equipped with such an assembly
10	10,729,842	Medical vial and injector assemblies and methods of use
11	10,729,828	Closed disposable multiple sterile blood bag system for fractionating blood with the corresponding method
12	10,729,722	Dialysis solution, formulated and stored in two parts, comprising phosphate
13	10,729,721	Dialysis solution, formulated and stored in two parts, comprising phosphate
14	10,729,673	Taxane particles and their use
15	10,729,666	Use of GABAA receptor reinforcing agent in preparation of sedative and anesthetic medicament
16	10,729,655	Orally disintegrating tablets
17	10,729,621	Acoustic reflectometry device in catheters
18	10,729,620	Baby bottle apparatus
19	10,729,619	Capsule sealing composition and its sealing method thereof
20	10,729,618	Liquid medicine filling device and liquid medicine filling method

FIGURE 4.13 CPC classification search results for class A61J.

able to simplify your design? Did you improve the robustness of your design? Did you reduce the need for specialized tools, devices, and software?

Typical X's that capstone design teams select are safety, cost, assembly, manufacture, and operability. Your team will choose your X values to best match the design project and the team member's skill sets. For each X value, you should conduct a literature search to find appropriate, relevant, and contemporary publications pertinent to your design project. The sponsor of the project can help prioritize the X values.

4.5 DESIGN SPECIFICATIONS

Design specifications or engineering design specifications are a set of constraints or criteria that the design must satisfy. Engineering design specifications must be quantitative and measurable with a range or specific target value and must have units.

Design parameters are specifications, requirements, or customer desires that are explicitly expressed in a quantitative form. For example, for the pill dispenser example, we may state that the device can hold five different pills, where the maximum dimension of any tablet is less than 1 in.

Defining the Design Problem

Designers should avoid including vague design specifications, which the customer may directly state. For example, the customer may indicate that they want the software to be user-friendly. How can we quantify user-friendliness? At the end of the design process, we should measure our design's performance against the design specification. If the specification is not quantitative (numeric), it will not be easy to measure and compare. Figure 4.14 shows the process for checking a particular design specification before including it.

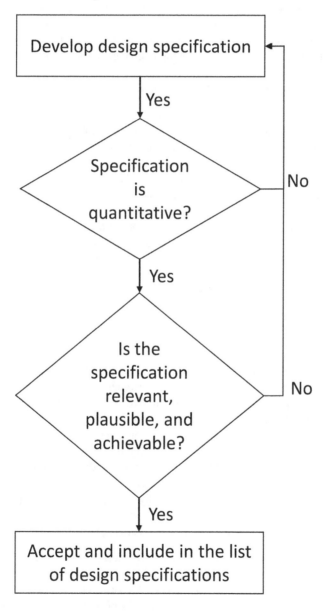

FIGURE 4.14 Check design specifications for correctness and relevance.

The numerical values associated with design specifications may be expressed as a range of insufficient information or in an early design phase. For example, we may require that the pill dispenser have a price of 300–500 when it is introduced into the market. The price range can be learned from surveys of potential customers for the product or by competitive analysis if similar products are sold in the market.

The design team will develop a deeper understanding of the design problem as the design process proceeds. Therefore, updating the design specifications is a necessary part of the design process. For example, suppose you encounter a design specification that cannot be met without disturbing the balance of the design. In that case, you should consider modifying the design spec in consultation with your project sponsor. Be prepared to explain why you are asking to change a design specification.

4.5.1 Customer Needs and Requirements

Capstone design projects generally have sponsors who are the customers for the design, or they represent a grouping of customers for the desired solution. The design problems are expressed in an open-ended form such that the design team must investigate and discover the list of needs or requirements. Finding the customer's needs and requirements can be challenging for the design team. The design team must translate the statements or definitions provided by the customer into a form that will be useful to the design process. Engineering designers should focus on stating the customer needs or requirements in the form of features or functions of the product and avoid including specifications on how the design performs that function or delivers that feature.

Customer needs or requirements (developed by the design team) should be expressed as clearly and accurately as possible. The team should share and solicit feedback from the sponsor on the statements. The statements should also be stated in a positive attribute of the final design product or process. For example, instead of saying that the product weight should not be too heavy, you can say that the product should be light and within some numerical range. Also, express the customer needs and requirements in some priority importance level instead of a hard constraint. You can work with the importance level to achieve more flexibility in meeting a plausible design. Figure 4.15 shows a translation of customer needs or requirements for the automated pill dispenser project into quantitative design specifications.

If the customer has many requirements, it will be necessary to group those requirements into logical groupings. Assign a relative importance value (say between 0 and 100) to each need and sort the list such that the most critical items appear first. By scanning the list, you should be able to eliminate redundant requirements. If you could group the requirements list into different objectives, you may need to move some items to balance and prune the hierarchy.

4.5.2 Design Specifications Leading Questions

In this section, we present a template to help the design team develop the specifications for a capstone design project. The questions presented and the answers to them can be stated as design specifications in a tabular form. This template will be helpful to both product and process-type design projects. Software design projects may need additional changes to this template (see Table 4.1).

Defining the Design Problem

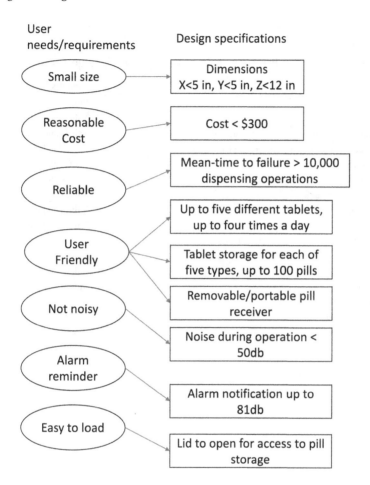

FIGURE 4.15 Translation of customer needs/requirements to design specifications.

4.5.3 Design Specifications Assignment

Use the template provided in this chapter to create an initial set of design specifications for your project. As your knowledge of the design problem and possible solutions matures, you will refine your design specifications further. Keep this in mind as you complete each stage of your design project, from many concepts to prototype to building your design solution.

TABLE 4.1
Working Template for Product or Process Capstone Design Projects

Parameters	Specific Values or Ranges
Name of project	Project name
Costs	Values and units
Functions – list each one if it can be quantified	Values and units
Performance parameters, e.g., acceleration, efficiency, power, reliability	Values and units
Installation parameters	Values and units
Operation parameters	Values and units
Maintenance parameters	Values and units
Safety parameters	Values and units
Dimensions	Values and units
Weight, load, pressure, stress, temperatures, humidity, etc.	Values and units
Special requirements such as a clean environment, etc.	Values and units
Environmental impact, sustainability, durability, reliability, etc.	Values and units
User parameters	Values and units
Standards and codes compliance	Numbers and references
Manufacturing specifications	Numbers and references
Materials specifications	References, values, and units

5 Conceiving Design Solutions

We educate and train engineering students to become excellent problem solvers. In every engineering course before the capstone design course, students are presented with mathematics, physics, chemistry, biology, and computer science knowledge that they learn and apply to practical close-ended problems in the textbooks or presented by the professors teaching those courses. Engineering students are very good at engaging in solution strategies when presented with such formulated problems.

Concept generation is a creative design problem-solving activity to collaboratively and systematically creating design solutions. Because design is an iterative process, so is generating solution concepts. Design solutions are based on ideas that the team members have formed about their design problem based on their efforts since project inception. Information from problem definition, literature search, competition research, patent search, and design specifications will guide the team members in generating solution ideas. The more concepts for viable solutions, the better. Having many plausible solutions will help the team achieve a more optimal design solution.

Creating design concepts is not an easy task. Engineering students have limited experience with a significant design problem, so their expertise is typically limited. The limited expertise may appear as a disadvantage initially, but it works out to be an advantage. Without a priori bias, students are not constrained and do not have a tunnel vision about a perceived solution to the design problem. Consequently, students think more creatively than seasoned engineers. Companies and industries are attracted to capstone design partnerships with universities because of the creative solutions that the students create during capstone design project work.

5.1 CREATIVE THINKING

The dictionary definition of creativity is "the ability to transcend traditional ideas, rules, patterns, relationships, or the like, and create meaningful new ideas, forms, methods, interpretations, etc.; originality, progressiveness, or imagination."

Creative thinking is essential for capstone design. But what does it mean to be creative? Do you know any creative people? Do you consider yourself to be creative? Can you learn to be creative?

Many researchers have studied creativity and creative thinking, and the search for methods and causes for creative thinking continues. While we do not fully understand what makes some people more creative than others, we recognize creativity when we encounter it. We see it as a brilliant idea or design that we have not seen before or previously thought of. When you conducted your patent search assignment, you may have noticed that some inventions were particularly creative.

You probably have had creative ideas that you can remember. When you encounter a new situation or new problem, can you think of ways to solve that problem or a workaround that seems creative to you?

5.1.1 Fostering Creative Thinking

While there is no formula or recipe for creative thinking, some behaviors or actions can facilitate original thoughts and ideas. The following techniques have been suggested by researchers of cognitive processes and creative thinking.

Silent Brainstorming – Brainstorming alone and silence can be helpful where original ideas can be thought without distractions. Whatever environment that makes one comfortable and relaxed can be beneficial to creative thinking. Some people do their most creative thinking while taking a shower! Group brainstorming dynamics are very different and can be an obstacle to creative thinking because of the processes such as taking turns or dominating personalities who push their ideas without respect for team members.

Attitude – Attitude is the most critical factor in success in capstone design and creative design. A positive attitude towards the project and fellow team members makes all of the difference between success and failure. A positive attitude and a genuine desire to solve the design problem will enable you to think creatively and contribute to the project. A positive mindset allows you to keep an open mind about your team members' ideas and encourage them to engage in the design project fully. A positive attitude also opens the imagination and inspiration towards a plausible solution.

Utilizing the Information Gathered – The information you have collected in literature search, patent search, surveys, and problem definition will provide a foundation for original thinking deriving new innovative ideas from those you have learned in your research about the design problem. The design specifications you have developed will set a reasonable scope for the design problem and the solutions.

5.1.2 Barriers to Creative Thinking

In parallel to engaging in methods that foster creative thinking, one should be mindful of some behaviors that block creative thinking. The following practices should be avoided or eliminated by communication, accountability, and transparency within the design team.

Poor Attitude – A negative attitude towards the project or the team will negatively impact the team's ability to work together and achieve an excellent design solution. A poor attitude can include stereotyping, minimizing the problem, showing minimal effort, taking a lone ranger attitude, expressing disrespect against others.

Tunnel Vision – Some design engineers get fixated on a particular solution strategy or method that limits their thinking and contribution to creativity and negatively influences team members in thinking creatively. When team members recognize this behavior, they should point it out to the team member and refrain from pushing only that solution.

Conceiving Design Solutions

Fear – Some students are not comfortable with the elements of capstone design, where they have to think outside of the zone of their comfort. They fear taking risks or entering a domain where their knowledge is uncertain. Some fear the open-ended nature of the capstone design problems where a solution is not guaranteed, or they are not sure which of the many possible solutions they should choose. They fear failure. Their fear can become a self-fulfilling prophecy. Team members should recognize this and work together to overcome fears or seek outside help if needed.

Poor Technique – Some student design engineers make poor choices in problem-solving strategies by not following the design steps carefully. For example, a poor problem definition or design specification set can be highly problematic for the design team. If the design problem is not defined well, it may create a situation where a solution cannot be achieved. If the design specifications are not fully developed or contain ill-conceived parameters or judgment, then a design solution may not be achievable.

Inadequate Domain Knowledge – Some design problems may involve aspects beyond the standard educational background of engineering students in a particular major. For example, a mechanical engineer may encounter a situation where they need to prevent electromagnetic interference in their design. They will need to seek expertise outside of their team to understand the science and work towards a solution.

Errors – Many different types of errors can cause frustration for the design team. One example of an error is incorrect assumptions or information provided by the sponsor, or perhaps a team member looked up wrong information. Errors can creep into the design process for many reasons, including lack of attention to detail. For example, the units on some numerical value could be transcribed incorrectly, or a calculation by one of the design team members could contain errors. Errors can be devastating to a design project. All engineering work, calculations, assumptions, data collection must be performed accurately and checked and double-checked.

As an example of engineering error, in 1999, NASA lost its Mars Climate Orbiter because the engineers failed to perform a unit conversion from engineering units to metric when communicating the critical spacecraft data before launch. The error cost NASA the $125 million spacecraft. The navigation team at NASA's Jet Propulsion Laboratory (JPL) used millimeters and meters in their calculations. At the same time, design engineers at Lockheed Martin Aeronautics in Denver provided acceleration data in feet and pounds. Lockheed Martin was responsible for the design and build of the orbiter spacecraft. JPL engineers assumed the acceleration data was provided to them in Newton-seconds. A simple error cost the mission years of setback, and much hard work was lost. SI units, also known as the metric system, are the international standard for communicating science, engineering, and technical data.

5.1.3 Techniques for Creative Thinking and Brainstorming

Some recent research in cognitive-based creative thinking (Ritter and Mostert, 2017) has shown that creative thinking skills can be improved using training based on established methods. We will introduce some of those techniques here.

5.1.3.1 Technique 1: Silent Brainstorming

Brainstorming is typically carried out as a group activity, but it can also be beneficial if carried out alone and individually. Brainstorming alone can be helpful for creative thinking because one can generate ideas without the boundaries, procedures, and restrictions present during group brainstorming. Brainstorming alone also eliminates the influence of views and criticisms expressed by team members during group sessions. An individual brainstorming session does not have to be very long. Five to ten minutes of silent brainstorming and writing down one's ideas is what is needed. Once unique ideas have been captured, they can be included in the group brainstorming session.

5.1.3.2 Technique 2: Evolutionary Thinking – The TRIZ Method

We will discuss the theory of inventive problem-solving created by Genrich Altshuller later in this chapter. TRIZ was developed to derive new design solutions from an analysis of previous design solutions. TRIZ was first focused on obtaining new patent ideas by studying prior patents. Analyzing some 50,000 patents, Altshuller and his coworkers suggested the "40 Inventive Principles." Using the TRIZ method, one can expand on the number of design ideas created with Technique 1.

5.1.3.3 Technique 3: Random Input

Lateral thinking is a way to escape from getting stuck in the same line of thinking for design solutions that can happen during brainstorming. Because sometimes our brains get fixated on solving the problem using the same methods repeatedly, some external input is needed to provoke our thinking into a different approach. The random input interrupts our line of thinking and points us into thinking differently, and therefore, the idea generation can become productive again. The random input does not have to be thoughtful but can be purely a random selection. For example, you can pick up a book and go to a random page, paragraph and pick a word. Then, try and relate the word to the problem at hand. A team member can also randomly think of a word, maybe from a song they like or a phrase they heard on the news or a friend. You are not limited to words in random input; pick an object or select a picture randomly and try to relate that to the design problem at hand. In relating the random input to your problem, you will be able to create new solution ideas.

5.1.3.4 Technique 4: SCAMPER

SCAMPER stands for substitute, combine, adapt, modify, put to other uses, eliminate, and rearrange. SCAMPER was introduced by Alex Osborne (1953), who developed a series of questions that spark ideas to diverge from a stuck path in brainstorming. Bob Eberle (2008) categorized those questions and created the mnemonic SCAMPER. The SCAMPER questions are designed to help generate more ideas during brainstorming by asking the questions in any order. Questions can be repeated, but the team members should take time to try and answer each question. Write down your answers for each team member for each question. When done, compile all answers to each question. Alternatively, you can use a whiteboard or large paper Tablet to record each team member's response to each question. Another effective

Conceiving Design Solutions

way to conduct the SCAMPER or brainstorming session is by using a shared file such as MScloud or Google docs.

- Substitute
  - What different processes can be used?
  - What different materials can be incorporated or substituted? What other sources of power could be applied?
  - What other applications can be substituted?
  - What users or groups can be included or excluded instead? What other methods can be used?
- Combine
  - What can be combined?
  - What purposes can be combined?
  - What groups of parts can be combined into one?
  - How can you combine parts for manufacturing or assembly? How can you integrate modules or sections?
  - How can you combine applications? How can you combine purposes?
- Adapt
  - What else is similar to this?
  - What other thoughts does this suggest? Does the past offer a similar situation? What can you adapt to other solutions?
- Modify
  - How can you add a new twist? How can you change the meaning?
  - How can you change the color or shape?
  - How can you change the geometry (expand or shrink)? How can you change the form?
  - How can you change the function? How can the motion be changed? How can something be added?
  - How can something be subtracted?
  - How can you increase or decrease the weight? How can you increase or decrease the load? How can you increase or decrease the force? How can you increase or decrease the pressure?
  - How can you increase or decrease the temperature? How can you increase or decrease the strength?
- Put to other uses
  - What other uses could it have?
  - How can it be put to other uses if changed?
- Eliminate
  - What can you eliminate?
  - What functions can you get rid of? What part can you remove?
  - What design specifications can you delete?
- Rearrange
  - What parts can you rearrange?
  - What other arrangements of the components will work?
  - What components can be rearranged or reversed? How can the process sequence be rearranged?

- How can the system layout be different? How can the cause and effect be different? How can the orientation be different?
- How can the methodology be reorganized?

There are many books and journal articles written on creative thinking. The bibliography lists many. You can find additional references by doing a literature search online at scholar.google.com, Academia.edu, and researchgate.net, or Amazon.com. The methods we have described here have proven to be sufficient for capstone design needs, but there is always room for more creative thinking in any design project.

5.2 GENERATING SOLUTION CONCEPTS

The methods discussed so far will help the design team members in generating many design concepts. How many concepts are enough? When does the team stop generating design solutions?

5.2.1 FUNCTIONAL DECOMPOSITION

Functional decomposition is also known as divide and conquer. The basic premise of the method is that you divide the large complex problem into sub-parts that you may solve directly. If a sub-part is still too complicated, then further division may be needed.

Functional decomposition is helpful for software design and development projects as well as process or product design projects. Functional decomposition can be performed at the stage of brainstorming. Instead of brainstorming solutions to the entire problem, you can brainstorm on parts resulting from the decomposition of the design problem.

Once the parts are solved, you can then assemble the solutions of the pieces into an overall design solution. Schematics help visualize the relationships of the components and in achieving decomposition. Each function in the decomposition will be something familiar to the designers where they either have created a solution to that problem previously or know how to obtain the solution from a reference source. Some functions will have a solution that is a part, a process, or a subroutine that can be acquired or purchased. As an example, Figure 5.1 shows the physical and functional decomposition of the SAE mini-Baja vehicle. Many engineering programs in the United States use the mini-Baja competition as one of their capstone design projects. Student teams design and build their mini-Baja cars and compete them during the spring in regional competitions. For additional information on SAE student design competition events, see https://www.sae.org/attend/student-events.

Functional decomposition is also a methodology used in systems design, where a system is decomposed into sub-systems, sub-sub-systems, and so on. At the end of each decomposition chain, a device or sub-system will be listed where a solution exists or is achievable.

Conceiving Design Solutions

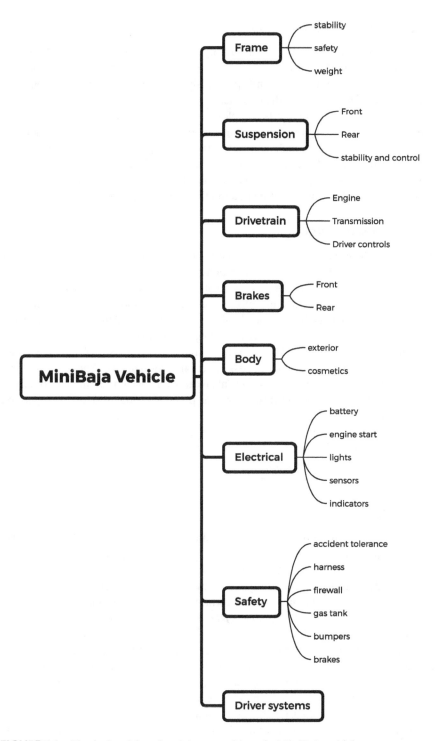

FIGURE 5.1 Physical and functional decomposition of a MiniBaja vehicle.

5.2.2 Morphological Methods

For product design, where a physical prototype or product is designed and created, the shape and form of the product or process may be a central factor. Morphological design is about shape and form and specifically about the arrangement of physical parts or steps of the design or the process.

Suppose the design problem is complex, a systems problem, or a process. In that case, it may be necessary to perform a decomposition before engaging in a morphological design solution for each sub-part of the design problem. Typically sketching or drawing may show the design solutions for each sub-part (or function).

The following drawings show a series of sketch solutions produced by students on a design team working on an automated intelligent pill dispenser solution.

5.2.3 Axiomatic Design

Dr. Nam P. Suh developed the axiomatic design methodology at MIT during the 1990s and further developed it in his book on Axiomatic Design in 2001. The design process described in this book is a derivative of axiomatic design. The two axioms in this method are the independence axiom and the information axiom. The independence axiom establishes that the functional requirements should be independent, and the information axiom minimizes the functional performance risk using an information content approach.

Nam P. Suh defined four classes in the design requirements hierarchy: customer domain (also known as customer needs), functional domain (also known as requirements or constraints), physical domain (also known as design parameters or specifications), and process domain (also known as process variables). He also suggested mathematical representations for the process, but that formalism is unnecessary for senior engineering capstone design. However, it is helpful for theoretical pursuits of design process formalism, automation, and artificial intelligence methods to improve design automation. The steps or stages that we have described in this book so far are derived from the axiomatic design process.

5.2.4 TRIZ Method

Russian engineer inventor, Genrich Altshuller, was interested in inventions. He believed that it was possible to create new designs by using information and features from past inventions. During World War II, he formulated the Theory of Inventive Problem-Solving or Russian – Teoriya Resheniya Izobreatatelskikh Zadatch (TRIZ). His line of thinking had many followers who further developed his ideas and many published articles appearing in *The TRIZ Journal* (triz-journal.com). Altshuller developed a set of principles for inventions from the initial theories and created a list of some 85 such principles. Many years later, in the 1970s, Altshuller (and others) further developed his theories into an algorithmic procedure as the forty principles of TRIZ. He termed the methodology as ARIZ (Russian acronym of – algorithm for solving inventive tasks) or algorithm for inventive problem-solving.

ARIZ (and TRIZ) defines ideality as a notion of measuring a good design solution. The greater the ideality, the better the design solution is. Ideality is mathematically defined as $I = B/(C+H)$, where, B = benefits, C = costs, and H = harms.

Conceiving Design Solutions

As B increases, I increases. As the sum C+H decreases, I increases. Benefits and harms represent the observed outputs of the design solution (called a system in the TRIZ/ARIZ method). Cost represents an input that the designers control. Benefits are desired outcomes of the designed product or process. Designers can control the benefits of how and what they design. The harms are the undesired outputs or features of the system.

For example, a designed system may achieve all of the desired specifications but may have very low efficiency and have a very high cost. Such an approach will have low to moderate ideality. By measuring and controlling the ideality during the design process, you can work more creatively and potentially achieve a more ideal (i.e., better) design.

The TRIZ and ARIZ methods contain a number of toolboxes that the TRIZ community has developed over time. Figure 5.2 shows the three toolboxes with their contents. We will briefly describe each in the following sections.

5.2.4.1 Contradictions

Identifying contradictions in a design project can help improve the design, increasing its ideality. Contradictions can be inherent to a system. For example, if you are trying to design a flying machine, you need more power in your engines to get more lift, but that will make the aircraft heavier where you want a lighter aircraft for better performance.

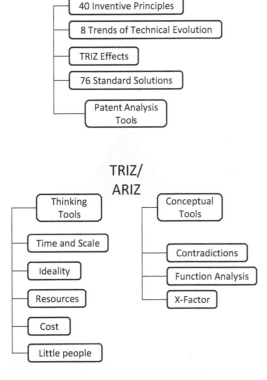

FIGURE 5.2 TRIZ and ARIZ toolboxes.

The TRIZ research on patents after World War II suggested that technical contradictions can be represented with 39 engineering parameters as:

1. Weight of moving object
2. Weight of stationary object
3. Length of moving object
4. Length of stationary object
5. Area of moving object
6. Area of stationary object
7. Volume of moving object
8. Volume of stationary object
9. Speed
10. Force
11. Stress or pressure
12. Shape
13. Stability of the object's composition
14. Strength
15. Duration of action by a moving object
16. Duration of action by a stationary object
17. Temperature
18. Illumination intensity
19. Use of energy by moving object
20. Use of energy by stationary object
21. Power
22. Loss of energy
23. Loss of substance
24. Loss of information
25. Loss of time
26. Quantity of substance/the matter
27. Reliability
28. Measurement accuracy
29. Manufacturing precision
30. External harm affects the object
31. Object-generated harmful factors
32. Ease of manufacture
33. Ease of operation
34. Ease of repair
35. Adaptability or versatility
36. Device complexity
37. Difficulty of detecting and measuring
38. Extent of automation
39. Productivity

Example – Consider designing a better filter basket for a swimming pool skimmer. Suppose your design is better in providing a better handle for removing it from the pool skimmer, and it increases the flow area of the filter service, and it would cost less because it will be easier to manufacture.

Conceiving Design Solutions

The contradictions in this new design can be identified from the list as 6 (area), 12 (shape), 20 (use of energy, i.e., flow resistance), 32 (ease of manufacture), 33 (ease of operation), and 39 (productivity). These factors can be analyzed for the new design. Does the new design solution improve on some attributes and make some others worse? Testing and experiments may need to be conducted to know for sure if some feature has improved and by how much. Alternatively, simulation and modeling may be performed to compare the different solutions. Regardless, a rigorous method should be developed or adopted to compare the alternatives and arrive at a quantitative difference between the TRIZ contradictions.

Resolution of contradictions will require the use of the 40 inventive principles as follows:

1. Segmentation
2. Taking out
3. Local quality
4. Asymmetry
5. Merging
6. Multi-function
7. Nested doll
8. Counterweight
9. Prior counteraction
10. Prior action
11. Cushion in advance
12. Equal potential
13. The other way round
14. Spheres and curves
15. Dynamism
16. Partial or excessive action
17. Another dimension
18. Mechanical vibration
19. Periodic action
20. Continuous useful action
21. Rushing through
22. Blessing in disguise
23. Feedback
24. Intermediary
25. Self-service
26. Copying
27. Cheap short-living objects
28. Replace mechanical system
29. Pneumatics and hydraulics
30. Flexible membranes and thin films
31. Porous materials
32. Color change
33. Uniform material
34. Discarding and recovering

35. Parameter change
36. Phase changes
37. Thermal expansion
38. Boosted interactions
39. Inert atmosphere
40. Composite materials

Many of these keywords are self-explanatory and should be familiar to most engineering students and professionals.

5.3 ANALYZING CONCEPTS

The list of viable concepts can be reduced and streamlined by an initial top-level simple analysis. This technique is called a go/no-go analysis. Each idea is reviewed, and the team decides which ones are not to be developed further and which ones show promise for further development. Perhaps 50% of the initial concepts can be eliminated from consideration at this point. The remaining concepts that show promise (i.e., received a favorable vote of most team members) for further development need to be analyzed in more detail. Methods of simple engineering analysis are highly discipline- and design-domain-dependent. For most designs, the analysis methods come from several areas of engineering and the sciences. The simple analysis method is typically based on simple formulas and correlations that allow the engineering team members to perform preliminary calculations and develop models to compare the concepts and decide which ones to keep. Because design problems are typically complex and have many competing requirements, analysis methods such as the Pugh matrix and the Quality Function Deployment (QFD) are used. At the end of the analysis process, the design team should have narrowed their concepts to two or three design concepts to develop a final design solution.

5.3.1 ENGINEERING ANALYSIS

Engineering analysis consists of applying scientific principles (chemistry, physics, biology) and mathematics and statistics to model, calculate, and predict the design's performance under varying assumptions and values for state parameters. Analysis methods range significantly in sophistication and detail. Preliminary analysis of design concepts is typically performed with simple models. Sometimes this hand calculation and schematics analysis technique is known as "back of the envelope" calculations. Such analysis methods are covered in the engineering science courses or discipline-dependent engineering courses and professional electives, as "example" problems or homework assignments. Typically, the hand calculations are also performed for a snapshot or steady-state system.

As the design advances, more sophisticated analysis methods may include complete systems analysis, systems codes, finite element analysis, nonlinear analysis, etc. Engineering computer software is generally available for many different physics or multiphysics analysis systems such as ANSYS, COMSOL, SIMULIA, MATLAB. Each software has capabilities for structural, thermal, electrical, electromagnetic,

Conceiving Design Solutions

chemical, fluids analysis, and many other specialized models. The licensing fees for such software are substantial even with the academic discounts, so the software to use will depend on the institutional resources to acquire and maintain the particular software system.

Advanced engineering analysis includes transient and dynamic calculations to simulate more realistic conditions. The use of advanced engineering analysis codes requires training and education so that appropriate models can be developed.

5.3.2 Modeling

Modeling can be computational or experimental. In some cases, it is beneficial to create a toy model of the design so that form and function can be better understood. Suppose the design solution is a product where a physical model can be built. In that case, modeling can be achieved by creating a smaller version of the design and experimenting with the prototype. 3D printers can help create a physical toy model to test and analyze design ideas. In some cases, the final design solution will be 3D-printed, so it is possible to build appropriate revisions to the initial 3D print model to optimize the final version.

If the design solution is for a process or a system that cannot be physically modeled, then a computational model will be the preferred solution. Computational models are input to software programs (computer codes) to create the geometry and associated boundary and initial conditions to simulate the performance of an existing or designed system for verification, validation, benchmarking, certification, parametric design, and insight.

5.3.3 Simulation

Simulation is the method of creating digital prototypes of the design to analyze how it works and predict performance in the real world. Simulation may be based on physics/chemistry/biology mathematical formulations, or it may be stochastic, where sampling is used to analyze the performance of a complex system.

The stochastic simulation includes critical algorithms such as Monte Carlo simulation and Markov decision tree analysis. Such statistical simulations can be developed in MATLAB®, Python, and programming languages such as FORTRAN, Visual Basic, C, C++, or Excel. Additionally, for specific disciplines and particular stochastic event simulations, computer codes such as ProModel can be used.

5.4 DECISION-MAKING

Decision-making is the process of choosing among two or more solution alternatives to find a practical design solution. Engineering decision-making is critical to the entire design process. Consider, for instance, in preparation for the critical design review; each design concept must be considered and compared to all of the other concepts. In fact, every step in the design process involves decision-making. The methodology used in decision-making directly affects the outcome and quality of the design solution.

Problem-solving and decision-making are related but are not synonymous. When a design solution does not meet its established design specifications or does not produce the expected or desired results or function as planned, it creates a problem that requires a solution. In general, decision-making can be considered a step in the problem-solving process. There have been numerous attempts to describe the human decision-making process. Based on education, training, and life experiences, each individual develops and employs a different set of logically ordered steps to identify and solve problems and make decisions. A formalized process and method must be utilized for problem-solving and decision-making in engineering design. Several tools for decision-making in engineering design are described next.

5.4.1 Pugh Chart

Pugh Analysis was invented by Professor Stuart Pugh, University of Strathclyde, in Glasgow, Scotland. Pugh Analysis is an evaluation matrix where alternatives or design solutions are listed in the first row, and evaluation criteria are listed in the first column. The method evaluates and prioritizes the many design solutions by scoring each design against each evaluation criterion.

The steps of a Pugh Analysis are:

- Develop the design specifications, which become the evaluation criteria.
- Generate design solution concepts (see the previous discussion on concept generation).
- Using a simple spreadsheet, list the evaluation criteria in the first column and the design concept reference numbers across the first row.
- Select a reference design, which may be one of the design concepts at random, or pick one that the team agrees is the one they like the best from intuition.
- For each of the evaluation criteria (design specification), evaluate each of the design concepts against the reference design, using scores of worse (−1), same (0), or better (+1).
- Record the rating in the corresponding row and column cell. Sum the results and select the alternative with the highest score.
- If none of the design concepts has emerged as the clear best idea with the team's consensus, look for a hybrid new concept that captures the competing alternatives' best features and rescore.

Figure 5.3 shows an example of a Pugh worksheet for an underwater jet scooter. The design solution concepts are numbered 1 through 30 on the top row. The evaluation criteria (design specifications) are listed in the first column. Each concept is evaluated with respect to case 1, and the results are recorded in the corresponding row and column. After evaluating all concepts against all criteria, sum the number of pluses and minuses. Repeat the analysis by selecting the highest value for the count of plusses and the lowest count of minuses for the same. In this example, design concepts 6, 13, and 14 have the highest count of pluses (6), and the lowest number of minuses is (3) for each of the three concepts. The team decides which of the three top concepts is used as the reference design in the next iteration.

Conceiving Design Solutions

Criteria	1	2	3	4	5	6	7	8	9	10	11	12	13	14	15	16	17	18	19	20	21	22	23	24	25	26	27	28	29	30
Drag	S	S	S	-	S	+	-	+	-	+	-	-	-	-	+	-	S	S	S	S	S	-	+	+	-	S	S	-	-	S
Safety	-	-	+	S	+	+	S	S	S	-	+	+	+	+	S	S	S	S	S	+	S	+	-	S	+	+	-	+	-	+
Hydrodynamic	S	S	S	-	S	-	+	+	-	+	-	+	+	+	+	-	-	-	+	+	-	S	S	+	-	-	S	+	S	S
Manufacturability	+	+	-	+	+	+	+	-	+	-	+	-	+	+	-	+	-	S	S	-	+	-	-	S	-	+	+	-
Cost	S	S	S	+	+	+	-	+	-	S	S	-	-	-	S	S	S	S	S	S	S	-	+	-	-	-	S	-	S	-
Durability	+	+	-	-	-	-	+	-	+	-	+	+	+	+	S	S	S	S	S	+	-	-	+	-	-	+	-	S	S	S
Weight	S	S	S	+	+	-	+	-	+	-	S	S	-	-	S	S	S	S	S	S	S	-	S	S	S	-	S	-
Functionality	-	+	-	-	-	+	S	S	S	+	-	+	+	+	+	-	-	-	+	+	+	-	+	+	-	+	+	+	+
Aesthetic	S	S	S	S	-	S	S	S	+	-	+	+	+	+	-	-	-	+	+	+	-	+	+	S	S	S	+	-	+
# of pluses (+)	2	3	1	2	4	6	3	4	2	4	3	5	6	6	4	1	0	1	3	5	2	2	5	4	2	3	1	5	2	3
# of minuses (-)	2	1	3	4	2	3	3	2	4	3	4	4	3	3	1	4	4	3	1	0	2	5	2	4	5	2	3	3	3	3

FIGURE 5.3 Example of a Pugh Analysis Worksheet.

This evaluation matrix is also known as a decision matrix. The decision matrix method can be used to evaluate and prioritize any list of options and, as such, is a decision-making tool. The criteria can also be weighted to achieve a more precise decision method. The weighted criteria method is a step in the QFD method described in the next section.

5.4.2 Quality Function Deployment

QFD is a structured method for evaluating, comparing alternatives, and making decisions in the design process. QFD enables the design team to specify the sponsor's or customer's wants and needs and evaluate each stated product or process feature concerning the impact on the design.

Quality in QFD signifies excellence as a goal. Function signifies how the design team will meet customer requirements and design specifications or how the product or process will function to meet them. Deployment defines how the design team will manage the design and development efforts to meet customer requirements and design specifications.

The first step in the process is to assess customer needs and requirements. As part of the design process problem definition, the design team has gathered information and created a list of customer or sponsor requirements (Figure 5.4).

Once the customer requirements (or needs) and their relative weights have been determined, then the process of constructing a QFD analysis diagram can begin.

Figure 5.5 shows an overview of the QFD diagram. The customer requirements and needs and their importance weighting factors are listed on the left-hand side of the chart. The engineering design team must create the engineering requirements (derived from the sponsor requirements). The engineering requirements are listed in the top row. Engineering requirements are quantitative requirements and typically have units associated with them. The matrix at the center of the chart is the relationship matrix, where a judgment is made about the relationship of the engineering requirements. Each cell is considered for a relationship value between the corresponding sponsor (or customer) requirement and the engineering characteristic. An importance ranking is assigned to each engineering characteristic (requirement) and entered on the bottom section of the chart. A total score is calculated for each engineering characteristic, which is then used to prioritize the different requirements. The roof of the chart is where information on the interrelationships of the engineering

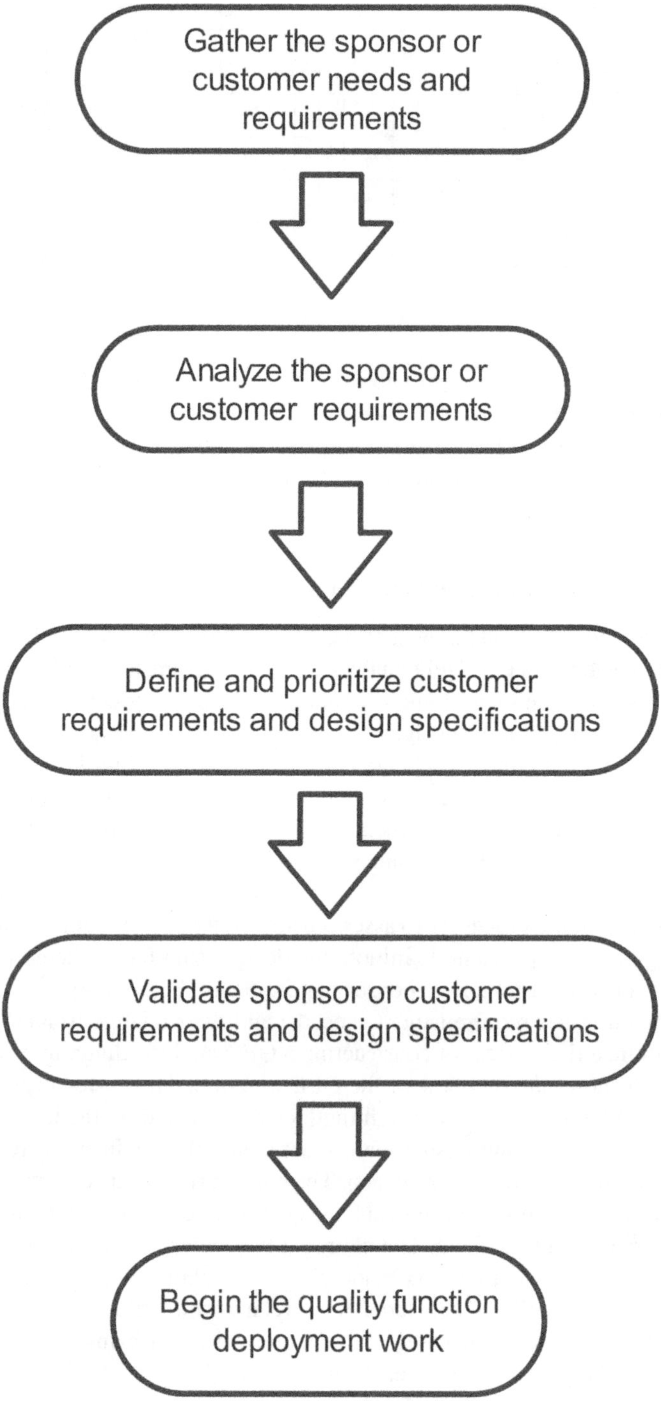

```
┌─────────────────────────────┐
│   Gather the sponsor or     │
│   customer needs and        │
│   requirements              │
└─────────────────────────────┘
              │
              ▼
┌─────────────────────────────┐
│   Analyze the sponsor or    │
│   customer requirements     │
└─────────────────────────────┘
              │
              ▼
┌─────────────────────────────────────┐
│  Define and prioritize customer     │
│  requirements and design specifications │
└─────────────────────────────────────┘
              │
              ▼
┌─────────────────────────────────────┐
│  Validate sponsor or customer       │
│  requirements and design specifications │
└─────────────────────────────────────┘
              │
              ▼
┌─────────────────────────────┐
│  Begin the quality function │
│  deployment work            │
└─────────────────────────────┘
```

FIGURE 5.4 Assess customer needs and requirements.

Conceiving Design Solutions

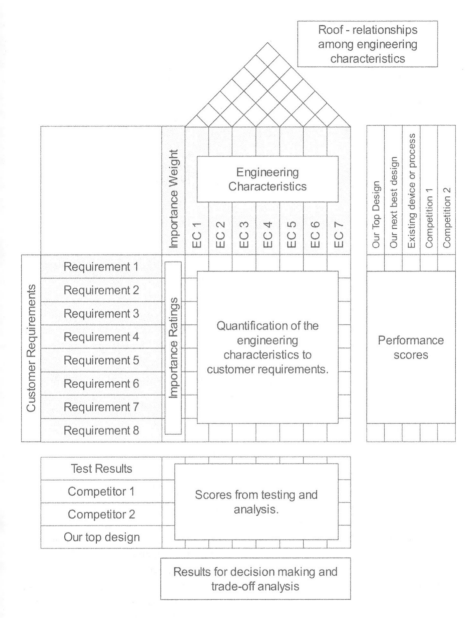

FIGURE 5.5 QFD overview.

criteria is recorded. The right-hand side of the chart contains a competitive analysis by entering comparative rankings for each design solution for sponsor (or customer) requirements—the relative ranking score ranges from 1 to 5, where 5 is best.

Figure 5.6 shows the entire cycle of a QFD analysis. The initial QFD analysis is for sponsor (or customer) requirements. The engineering criteria from the first analysis for the product become the requirements for the design (on the left-hand side),

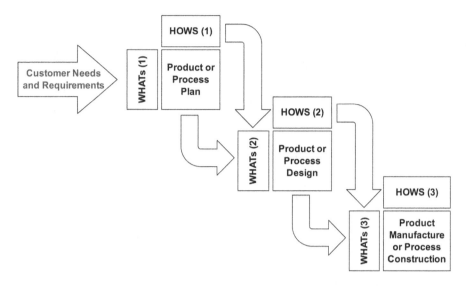

FIGURE 5.6 Complete QFD process for a product or process design and development.

and the QFD procedure is repeated for this stage. The design specifications, in this case, are entered at the top row. At the last stage of QFD, the design specifications are entered in the left column.

To illustrate the methodology and analysis results from the QFD method, we will illustrate the entire process with an example.

5.4.3 QFD Analysis Example

Consider the following problem: design and build an autonomous pool skimmer. The project skimmer must clean a 30′ by 50′ pool in 2 hours or less. The skimmer also must be autonomous, meaning that it cleans the pool surface without any user input. As an option, the pool skimmer should be controllable remotely. The skimmer parts must be 3D-printed with PLA. The budget for the project is $200. Optionally, the design team can consider charging batteries on the skimmer using solar panels or plug-in power.

The design specifications developed for this design problem are shown in Table 5.1.

The QFD is a powerful tool for deciding which design concepts are strongest compared with the design team's various designs and current skimmers on the market. The team of four students generated 120 design concepts and used the Pugh Analysis method to select three top designs to consider further. Furthermore, three of the more popular skimmers on the market were chosen for comparison. The team created a list of customer demands and engineering characteristics. From the problem definition and the design specifications, a list of customer requirements and engineering characteristics was developed by the design team as follows (Figure 5.7).

The customer demanded qualities for the skimmer are:

- Speed of a complete cycle of operation
- Light in weight

Conceiving Design Solutions

TABLE 5.1
Design Specifications

Length	18–22 in.
Width	18–22 in.
Height	4–10 in.
Weight	< 10 lbs
Catch area	14–16 in.´ 2–6 in. (28–96 in.²)
Waste capacity (volume)	400–1,000 in.³
Cost	< $200
Noise level	About 65 decibels
Speed of cleaning	Less than 2 hours to clean 30´´50´ pool on one pass (12.5 ft/min)
Battery Life	3 hours
The lifespan of robot	5 years
Warranty	2 years
Waterproofing	XTC 3D (Waterproof Material)
Autonomy Raspberry Pi	Arduino Uno
Buoyancy weight	< water weight displaced
Motors	DC, Servo, Stepper
Material	3D-printed ABS or Tough PLA
UV protection	Ultraviolet-resistant paint
Target market	Homeowners that have pools, commercial pools
The environment of use	Wet, Humid, Submerged, Sunlight
Market demand	10,000 units
Competing products	Skimdevil, Solar Breeze, SolaSkimmer, Skimbot, BETTA
Manufacturing	3D-printed
Time to complete project	May 2021

- Easy to use
- Portable
- Durable
- Quiet operation
- Battery life
- Easy to set up
- Adorable

The engineering characteristics for the skimmer are as follows:

- The maximum weight to be under 10 lbs
- The length and width to be 18–22 in.
- The height to be 4–10 in.
- Operating noise level below 65 decibels
- The catch area for debris to be 28–96 in.²
- The waste volume to be in the range of 400–1,000 in.³
- The number of parts to be less than 5

- The battery life to be 3 hours or more
- The market price of the skimmer to be less than $200
- The set-up time for the skimmer to be less than 15 minutes out of the box

Figure 5.8 shows the QFD chart after entering the customer requirements and engineering criteria. The next step is to assign relative weights to the engineering criteria and customer requirements. The magnitude of the weights is determined from team interpretation of the problem definition and customer requirements. The magnitude of the weights is based on a team consensus of the relative values of the weights. The weights will need reevaluation as the team develops a deeper understanding and intuition of the problem and the solutions emerging from the team's efforts (Figures 5.9 and 5.10).

Each engineering characteristic is compared to each customer requirement. Again, the assignment of values is based on the team's consensus. A strong correlation is assigned a value of 9, a moderate correlation is assigned a 3, a weak correlation is assigned a 1, and if there is no correlation, the cell is left blank (Figure 5.11).

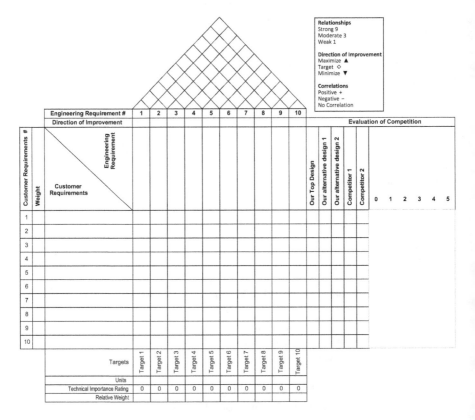

FIGURE 5.7 Blank QFD chart.

Conceiving Design Solutions

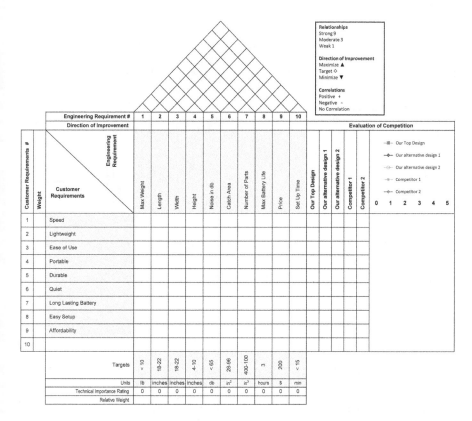

FIGURE 5.8 QFD chart after entering customer requirements and engineering characteristics.

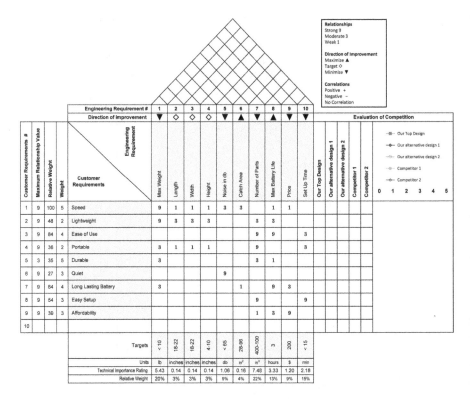

FIGURE 5.9 QFD chart after entering direction of the importance of engineering criteria.

Conceiving Design Solutions

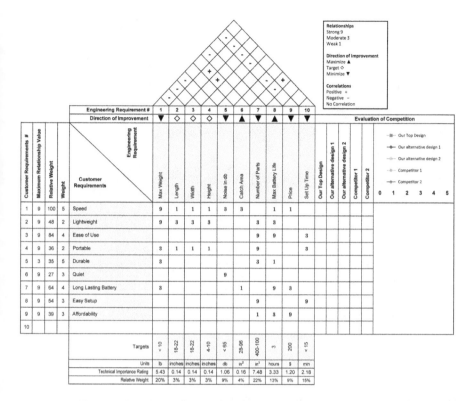

FIGURE 5.10 QFD chart after analysis of engineering characteristics correlations and entering the values.

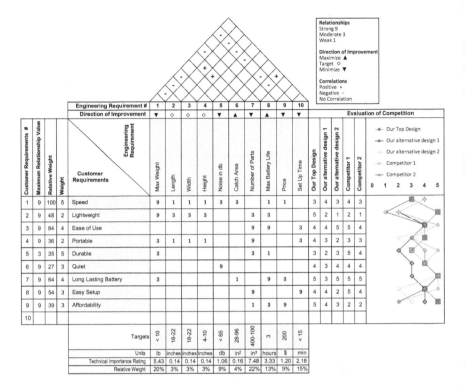

FIGURE 5.11 Completed first draft of QFD chart with an evaluation of competition.

6 Critical Design Review

A critical design review (CDR) is a technical review to ensure that the design project team can select a superior design idea before prototyping. The entire capstone design class, mentors, sponsors, and experts are the technical audience for evaluating the team's design work and top concepts. The student design team presents their top design concepts (typical two to four top concepts from the Quality Function Deployment (QFD) analysis) in a short but concise presentation (usually 15 minutes). A successful presentation CDR is centered upon demonstrating that the team's design concepts satisfy the design specifications and solving the design problem. The CDR is typically scheduled in a capstone design class after the team has completed the Pugh and QFD analyses. The capstone CDR mimics the process used in the industry to evaluate design concepts before they are prototyped or manufactured. However, it is adjusted in timing to meet the constraints of the academic semester (or quarter).

The capstone CDR process is based on a formal, technical, and concise presentation of the team's work and feedback and critique from the audience. The guidelines for what should be included in the presentations are described in the next section.

6.1 ENGINEERING TEAM PROJECT PRESENTATION

The CDR presentation should include critical topics to inform the audience and communicate crucial technical analysis and design steps.

- Include the definition of the problem.
- Include design specifications and their relative significance.
- Describe the basis for reducing the many design solution concepts down to two to four concepts presented during the CDR.
- Present the technical details of each of your solution concepts and how it matches up with the sponsor requirements
- Present your QFD analysis and explain the assessment of each of your design concepts and their comparison to each other and the competitive solutions.
- Allow sufficient time for the audience to ask questions and critique your design solutions.
- Take notes on the questions asked and comments offered by the audience.

6.2 RECEIVING CRITIQUE

The audience for the CDR consists of fellow students in the class, sponsors, mentors, teaching assistants, and professors. One purpose of the CDR is to solicit helpful suggestions and critiques from the audience. The audience's comments and suggestions can happen during or at the end of the presentation if there is sufficient time. A small class (<40 students) can allow time for interactive comments. However, it won't be easy for a large class (>60 students) to schedule sufficient time for each team to present their work and interact with the audience. If the presentation time can be scheduled for interactive comments and suggestions by the audience, then the design team presenting their work should capture the comments and questions. One or two team members should volunteer to be the recorders of questions, comments, and suggestions from the audience. Design project team members should provide their e-mail contact to the audience and ask that comments be e-mailed after the presentation.

In the case of large classes, teams may only have 10–15 minutes to make their presentation and engage in answering one or two questions and answers. Therefore, the professor for the class should invite comments either through a class survey instrument (a form created in Google Forms or similar software) that all members of the class should complete. The critique survey form is a more formal way to collect comments and suggestions for each team, and it can be a part of the grading for this component of the class. The comments for each team are provided to them after all comments and suggestions have been collected.

6.3 INCORPORATING FEEDBACK

Design teams should carefully review all questions, answers, comments, and suggestions they receive during and after the CDR presentations. Comments and suggestions should be sorted and categorized to assess their applicability and relevance to the project. The team should decide which comments or suggestions can actually improve the design and be implemented to change the concept or parts of it.

Making changes to the design concept will require deliberation among the team members and the sponsor to assure agreement regarding changes. The team should discuss what was learned during the CDR at their regular meeting after the date of the CDR presentation and invite their sponsor to participate in the discussion.

All relevant comments and suggestions should be referenced to the person who commented or suggested and documented in the design report as a citation. Sometimes it is necessary to meet with the person who made a significant and design-changing comment or suggestion to discuss their suggestion more in-depth. The team should follow up on all significant comments and suggestions.

7 Proof of Concept

Creating a fresh, brilliant solution concept for a design problem is an exciting venture for the student teams. However, before building and testing their solution ideas, they must pace their design and developmental process and validate their solution ideas. Therefore, the team needs to have a comprehensive project plan that includes a proof of concept (POC) design. Creating a POC can mitigate errors and overlooking critical aspects of their project that would cause them to fail.

Moreover, suppose the team wants their sponsor to accept their ideas and design solution proposals. In that case, project teams need to prove that their ideas are practical, functional, viable, and worth the additional human and financial resources needed to complete the project.

A POC is a crucial step for capstone design projects. A POC is manifested by presenting the proposed design solution and its potential viability for further development.

The team describes their top ideas and proposed functionality, including specific design features and their practicality. Then, key design features are presented and proved through a prototype of those design features. The POC presentation and approval are prerequisites before the team proceeds to build their design solution.

A POC typically involves a small-scale visualization or prototype to verify the real-life functioning of the idea. It's not yet about delivering that design concept but demonstrating and proving its feasibility. Through the POC, the design team can prove that building the proposed solution, product, process, or method is achievable. POC also allows the sponsor to see and understand the idea's potential, giving them a glimpse of what the team intends to develop further. In this way, the design team can ensure their solution supports the sponsors' needs and the overall requirements of the capstone course.

Designers often use POCs and prototypes interchangeably. A POC shows whether the product or process can be built or not, while a prototype physically presents its essential functions. When design teams prove that the design solution is valid and sponsors agree and approve the proposal, they can build the product or create the process. The POC aims to validate the idea or assumption, and prototyping lets the design team realize the concept by creating an interactive working model of their proposed design solution.

The design prototype has the functions, operating features, layout, aesthetics, and other components of the solution captured by POC. The prototype proves the usability of the proposed solution. It doesn't have to be perfect, similar to the POC ideas. Prototyping is an attempt to develop and test the critical features of the proposed solution. It demonstrates how the design solution works as provided in the POC.

Having a POC and prototype is essential for student design teams to refine their ideas and begin their product or process development process. In addition, the POC helps the design team identify unknowns and obstacles they may face in further

DOI: 10.1201/9781003108214-8

developing the proposed solution. Uncovering the obstacles and unresolved issues during the POC phase mitigates problems later during the build and testing of the design solution, reducing the probability of project failure. The POC does not guarantee the smooth progression of the project, but it does increase the project's success. The design team, mentors, and sponsors can find ways to eliminate, mitigate, and address the unresolved issues and unknowns and realize their project's success.

Design projects for capstone have a limited budget, so it is not feasible to create an actual product in many cases. Instead, the build will be a scaled-down version of the physical product or process. The scaling is necessary to create a practical final design solution. Therefore, the POC and prototyping provide information and the basis for the feasibility of the design solution and its scalability. In addition, POC can help the design team and the sponsor plan for scaling for manufacturability, production scaling, scoping, or process implementation.

Before the design team can be assigned resources for their project implementation, they should demonstrate that the expenditure of resources will be worth it. POC is the design team's opportunity to demonstrate the viability of their methodology and design solution. They can show the design solution idea in detail with illustrations, visuals, and prototypes to provide the sponsor with sufficient information to approve moving forward to build and test.

Suppose the POC is not successful (i.e., no proof of design is achieved), or the prototype does not verify the design solution. In that case, the design team must reconsider their choice of a design solution and strategy. The design process cannot continue until the team has achieved a POC.

7.1 HOW TO CREATE A PROOF OF CONCEPT

The POC process has several basic steps that design project teams should follow.

7.1.1 DEMONSTRATE THE DESIGN SOLUTION MEETS SPONSOR REQUIREMENTS

The POC should establish the need for the product or process. As discussed previously, market surveys, sponsor reviews, focus groups, and end-user interviews are some methods to establish the need for the proposed design solution and features.

The design team should develop and ask thoughtful questions about the sponsors' (or customers') likes, dislikes, and desires of their proposed solution. Sponsors' feelings, intuitions, and perspectives should be documented and be part of planning the progression of the design work. Having that information will guide the design team in streamlining the design solution and fine-tuning the design specifications.

7.1.2 GENERATE AN IMPROVED DESIGN SOLUTION

The results of the interviews with the sponsors, potential users, customers can help the team brainstorm improvements to their original design concepts. Finally, improved designs should be analyzed by incorporation into the Quality Function Deployment (QFD) analysis and processed for new ratings.

The QFD chart's sponsors' needs and requirements and the engineering criteria may need to be revised to comply with the new information collected from the interviews.

7.1.3 Build a Prototype and Test It

Once the design team has generated a feasible idea from the interviews and updated QFD, they should create a prototype based on the decided requirements, criteria, design specifications, features, and design solutions.

The project team must have the sponsors and individuals in their interview groups experiment with the prototype solution. The trials assist the design team in verifying and validating their design choices. In addition, testing the updated design with the same individuals enables the design team to document their feedback before and after prototyping.

Building a prototype should be quick and efficient. It does not have to be complete or perfect. The prototype can be a scaled-down version, perhaps 3D-printed or quickly modeled in a process design software. The student team will have a limited time to work on the prototype for POC, perhaps a month, so the team will need to act quickly but be precise and diligent.

7.1.4 Collect Test Data, Analyze, and Document

The design team must collect data and document the test group feedback concerning their experiences, satisfaction, reactions, complaints, and other pertinent details. The analysis of data collected from testing and sponsor/customer feedback allows the design team to validate their design and verify the feasibility of the solution. It also provides insight for the team to adjust their project planning moving forward.

The information gathered from testing, interviews, and analysis of that data and actions taken should be entered into the team's project management plan.

7.1.5 POC Presentation

The POC presentation is the culmination of the first phase of a capstone design project. In a two-semester course sequence, the POC step comes at the end of the first semester and before the team completes their preliminary design report. The POC presentation is a formal presentation where the design team presents their work to their student peers, sponsors, mentors, and professors.

The design team must present, among other things, the customer requirements, problems, and dislikes for the initial design that the POC design solution solves. The team is to present features that address those issues and design elements that demonstrate the benefits. They should describe the product or process development process and how they have managed the project and illustrate a Gantt chart of the layout.

They should include clearly defined success criteria, milestones, project management metrics, evaluation metrics, timelines, future plans, resources needed, and other aspects discussed earlier. Then, after the team persuasively presents their POC design, they will likely receive approval to build it and test it.

POC helps design teams and sponsors examine if a proposed design concept is practical and valuable for the target audience, and achievable. Design project teams can explore the planned features and functionalities of the solution along with the financial analysis, resources, and capacities required to implement it.

A POC is a design step to test the design solution and underlying assumptions. The primary purpose of the POC is to demonstrate the functionality and features. Prototyping is a necessary step in capstone design that allows the stakeholders to visualize how the product works. It is an interactive functional model of the end solution.

7.1.6 Viability and Usability

The usability of the POC in the real world is to make decisions about a project's probability of success and assess if the fundamental assumptions about the proposed design solutions are viable. The POC step is to identify the design solution features and functionality before spending lots of precious resources. The solution prototype is a first attempt at creating a working model that might be usable in the real world. Of course, many things can go wrong in the design process, but identifying wrong assumptions, mistakes, errors, unknowns, stumbling blocks, and unresolved issues is the principal purpose of building a prototype. A prototype should have most of the functionalities and performance of the end solution but will generally not be as efficient, aesthetic, durable, actual size, or materials.

The POC step facilitates sharing design solution knowledge among the design team members, investigates design techniques, and provides evidence of the validity of the solution concept to the sponsor. The prototype model extends the POC by developing a working model to visualize a snippet of the final design solution.

Prototyping is a quick and effective way of illustrating the value of the design solution to stakeholders. In addition, it is a platform to receive feedback and guidance from sponsors, mentors, and end-users. This methodology also helps in the documentation phase and provides the design team with a helpful estimation of resources and time needed for project completion. The final POC does not have to be bug-free but should ultimately show the functionality and performance of the solution concept.

8 Documentation

"Documentation" encompasses a broad scope and subject area. It encompasses writing, organizing, project management, critical thinking, analysis, and engineering problem-solving. Additional subtopics can be identified within each of the broad areas.

Documentation is a process encapsulated in rules, guidelines, and suggestions. The rules and guidelines must be clearly established in the capstone design class.

An informal or ad hoc approach to documentation significantly reduces the probability of the design project's success. Additionally, lessons learned from past designs will be lost if not captured in an organized manner. A formal set of guidelines for documentation will assist future capstone design teams in learning the capstone design process and having access to examples that will assist those teams in their projects.

An additional benefit of a formal documentation process is that students learn by doing, and following and creating the documentation will benefit them in their future professional endeavors. In addition, good documentation skills are essential for practicing engineers.

Finally, sponsors of capstone design projects demand and appreciate thorough documentation in the form of design reports, engineering presentations of intermediate results, progress reports, and regular communications. Satisfaction of sponsor expectations for documentation is essential to the success of an engineering capstone design program.

8.1 ENGINEERING DESIGN DOCUMENTATION

Documentation is a critical element of the engineering design process. Because engineering design is a complex process and involves many analysis and decision-making points based on data and evidence, all aspects must be captured and documented so that others can clearly follow and understand the basis for the design decisions and final results.

Documentation comes into play at each step in the engineering design process. Communication of ideas, progress, problems, unresolved issues, and solutions is essential to keep all stakeholders informed and sponsors appraised of the status of the design project. One of the tools for documentation is an engineering logbook. The logbook is a personally written recording of the evolution of ideas, problems noted, drawings, schematics, analysis, data, and calculations, similar to a project's diary.

Meeting minutes are another form of documentation, as discussed previously. Meeting minutes should capture design discussions and incremental work performed on the design project. In addition, meeting minutes should document design steps and critical decisions.

During the capstone design cycle, the design team will collect and generate a significant amount of information. This information must be cataloged and organized. A proven way to keep the information organized is in a design binder. Alternatively, the same information can be kept in electronic form. Any paper documents should be scanned and included in the design binder folder hierarchy.

Design presentation slides are another form of documentation. The presentations include critical information on the critical design review, proof of concept, and building and testing the design.

The first comprehensive document is a preliminary design report (PDR). The PDR captures the details of the design process leading to proof of concept and a prototype. It serves as a significant milestone in the project to communicate information to the sponsor, and it also serves as a culmination of the first phase of the design project.

A brochure and poster can also serve as an excellent method for communicating a summary of the project with a broader audience, including a conference or a design showcase.

Finally, the final design report (FDR) serves as the complete documentation of the entire design work of the team. Therefore, the FDR must capture all essential aspects of the design work of the team.

8.1.1 Engineering Logbook

Each design team member should maintain an activity journal of their design work and project contributions in an engineering logbook. In the engineering profession, a logbook is necessary to document the personal and intellectual work of the engineer in a way that it can clearly demonstrate their contributions and possible inventions. In addition, the engineer's logbook may be utilized in legal cases of proving intellectual property origins on patent cases.

The logbook should be a stitch-bound notebook (see Figure 8.1) with each page sequentially numbered so that pages cannot be added or removed after the logbook was bound. Logbooks can be purchased with pages formatted with entry fields for dating the page and signature box (Figure 8.2).

Each entry in the logbook should be dated and signed by the engineer. No pages should be left blank so that it is possible to clearly see the timeline evolution of ideas and progression of the design solution. Figure 8.3 shows sample pages from an actual logbook.

Effective engineers use the logbook as a way of documenting their ideas, writing down creative thoughts and ideas, sketching thoughts and concepts, mind mapping, and documenting successes and failures. As such, the engineering logbook becomes a learning workbook.

The engineer should record the date on each page and start a new page for a new date. Each entry should be labeled with a title and included in a Table of Contents for the logbook. A permanent ink pen should be used for the logbook entries. Entries should not be erased. Instead, an incorrect entry can be crossed out with a single line and initialed by the engineer. Pages in the logbook should not be left blank, skipped, or removed. If necessary to record an entry that was not included previously, it can be added with a date and reference to the page number where it is related. All activities, including successes and failures, should be included in the logbook. Sometimes failures are more informative than successes.

Documentation 115

FIGURE 8.1 Engineering logbook styles.

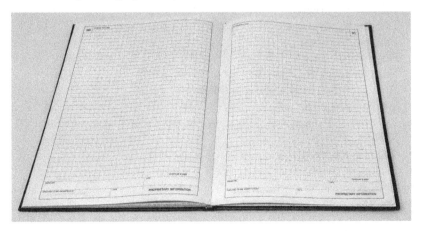

FIGURE 8.2 Engineering logbook blank page.

As a minimum, the design engineer should regularly enter the following into their logbook:

- Personal and team activity on the design project
- Summary of communications regarding the design project with team members, sponsors, mentors, and other stakeholders

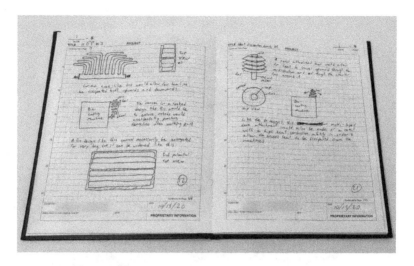

FIGURE 8.3 Engineering logbook sample page.

- Sketches of design concepts or ideas
- Engineering analysis of concepts and ideas, calculations, and formulations
- Questions, answers, and other ideas on the design project
- Summary of key decisions as a result of meetings or communications
- Sponsor needs/requirements
- Engineering criteria
- Design specifications
- Problem definition
- Design objectives
- Meeting notes, outcomes, and action items
- Brainstorming, results, and action Items
- Work-in-progress
- Keywords for gathering information, literature search, and patent search
- Sources of ideas
- Test engineering plans, parameters, and methods
- Evaluation of data and results
- Critical design review notes
- Decision matrix and criteria
- Design process steps and actions
- Project reflections
- Team performance

8.1.2 Design Binder

A design binder is an efficient way for the project team to collect and organize their project materials. For example, a typical engineering capstone project conducted over two terms would require a 3 or 4-in. three-ring binder to collect and store the relevant documents.

The design team should take their design binder to all critical meetings with the sponsors, mentors, and capstone professors. Having the design binder during project review meetings allows the reviewers to see an immediate snapshot of the entire project.

As a practical style guide for preparing the design binder, materials for inclusion in your design binder can be grouped under the following tabs:

- **Index and Table of Contents** – Provide a Table of Contents or index that includes the heading for each section.
- **Problem Definition** – Include problem definition and supporting materials for the problem being addressed as created by the team.
- **Team Work** – Include team meeting minutes, weekly progress reports, and e-mail correspondence with the sponsor.
- **Engineering Analysis** – Include notes, sketches, engineering analyses/calculations, drawings, charts, figures, and schematics.
- **Project Plan** – Include the project plan with baseline and updates, Gantt chart, and calendar of team tasks/schedule/deadlines with milestones of your project. The project plan must include resources allocated to each task, including people and equipment.
- **Presentations** – Include any PowerPoint presentations created by the team during the project duration.
- **Literature and Patent Searches** – Include results of literature and patent search exercises collected by all team members. In addition, include copies of all articles, sections of books, journal articles, and complete patent documents.
- **Design Approaches**
  - Include description and details of any design methods applied to your project (from this textbook).
  - QFD – Include the Quality Function Deployment analysis of your concepts and your evaluation of competition
  - CAD or hand-drawn sketches, blueprints, or plans of your design
- **Design Specifications** – Include team's work on creating design specifications and include revision history of design specifications with dates for each revision
- **Bill of Materials (BOM)** – Include BOM cross-referenced to drawings, supplier lists, and contact lists.
- **Systems Analysis** – Include relevant analyses of systems and computational analysis (such as the load-bearing capacity of something, strength, heat transfer analysis, materials behavior, process layout).
- **References** – Include competition rules, industry material or specifications, articles, secondary research, applicable standards (MILSPECs, ASTM, ASME, etc.). Include
  - Papers, journal articles, patents, reports, manuals, industry materials, etc.
  - Any other documents generated or used in the project
  - References to engineering standards (MILSPECS, ASTM, ASME, ISO, etc.)

- **Modeling** – Include any models, simulations such as CFD, dynamics, kinematics, stress-strain fields, using computer codes such as CAD, Abaqus, Fluent, and Comsol.
- **Trade-Off Analysis** – Include radar/spider charts, QFD models, or formal models for comparing alternatives and decision-making.
- **Financial Analysis** – Include
  - Cost analysis of the design (one copy versus many)
  - Personnel level of effort and cost estimation
  - Analysis of manufacturing costs and mass production
  - Use of facilities such as 3D printing or laboratories
  - Consultation time with professors and other consultants
- **Critical Thinking and Analysis** – Include analysis results concerning economic, environmental, social, political, ethical, health and safety, manufacturability, and sustainability.
- **Administrative** – Include administrative paperwork, including purchase orders, bid sheets, quotations, competition registration, and conference registration.
- **Resumes** – Include the most recent copy of resumes for all team members.

8.1.3 Electronic Files and Project Archive

Electronic files created during the design project are an essential part of the documentation of the design work. The electronic files for the design project should be organized and be available to all team members (including professors and teaching assistants) and possibly mentors and sponsors. A suggested organization of information is presented in the following list of folder names:

- Additional considerations
- Administrative
- Assessment
- Brochures
- CAD files
- Concepts
- Cost analysis
- Critical design review
- Design for X
- Design specifications
- Final design report
- Manuals
- Meeting minutes and notes
- Paper
- Patent search
- Photos
- Poster
- Preliminary design report
- Presentations

Documentation

- Previous project information
- Problem definition
- Project management
- QFD analysis
- References
- Resumes
- Testing
- Videos
- Weekly progress reports

The electronic files can be shared with cloud sharing file systems such as Google Drive or DropBox. In addition, many universities have standardized on the Google application suite (G Suite).

8.2 VERBAL PRESENTATION WITH SLIDES

Verbal team presentations in capstone design are common, and typically a student design team is expected to make two or three presentations during the period of an academic term (semester or quarter). Design presentations are typically scheduled for 10–30 minutes. Therefore, the presentations must be prepared carefully, concisely, and efficiently by the student team. They must also practice the presentation multiple times to assure a professional delivery within the allotted time. There is no set number of presentations. The team must practice as many times as necessary to achieve a perfect presentation to communicate the progress made, significant accomplishments, and future direction for the project.

Tips on preparing an excellent presentation are shown below in the form of a presentation on presentation tips (Figures 8.4–8.37).

8.2.1 Guidelines for Capstone Design Presentations

The design presentations should follow some basic guidelines to allow for all team members to participate and convey a professional, efficient, and informative project status to the audience. Typically, there is a limited time available because of class size and scheduled class meeting hours for the capstone design course. Therefore, the professor for the class should communicate the guidelines for the presentations, including a rubric for assessing the presentations. Student teams should use the guidelines and rubrics to fine-tune their presentation content and style of delivery.

Preview, or Coming Right Up . . .

- Tips on presentations
- Tips on PowerPoint/Keynote

FIGURE 8.4 Presentation tips outline.

Why Use a PowerPoint/Keynote Presentation?

- Presenters using visuals perceived as
  - Better prepared
  - More persuasive
  - More interesting
- Visuals can supplement/enhance message
- Visuals provide another means to learn and remember message

FIGURE 8.5   Why use PowerPoint?

It's Got the Look . . .

- Use 20-pt. as smallest font if possible
- Use white space – a document's best friend
- Consider *sans serif* fonts for easier read
  - Arial
  - Helvetica
  - Geneva

FIGURE 8.6   Presentation format – it's got the look 1.

It's Got the Look . . .

- Bullet at appropriate levels to convey meaning
  - Sub-heading
    - Sub-sub-heading (20 pt. font here—how does it "read"?)
      - Let's try a sub-sub-sub heading (this is 18 pt.)

FIGURE 8.7   Presentation format – it's got the look 2.

Be a (Wo)Man of Few Words

✦ One main point per page

✦ 6 x 6 rule (of thumb)
  ✦ 6 words per bullet
  ✦ 6 bullets per page

✦ Breaking up is hard to do, but do it

FIGURE 8.8 6×6 rule.

Be a (Wo)Man of Few Words

✦ Slides serve as brief outline, place markers, enhancers

FIGURE 8.9 Be brief.

Be a (Wo)Man of Few Words

✦ Use words or phrases, not sentences

✦ Delete articles (a, the) & less critical words

FIGURE 8.10 Be efficient.

Be a (Wo)Man of Few Slides

✦ Estimate 1 – 2 minutes per slide
  ✦ ~ 5 – 10 slides for 10 minutes
✦ But rehearse to check on
  ✦ Time
  ✦ Coordination
  ✦ Sufficient/appropriate phrasing

FIGURE 8.11 Timing.

General Tips on Oral Presentations

1. Define the problem or state the central question being addressed
2. Indicate its importance
3. Briefly tell what was done & how
4. State what was found
5. Consider broader implications of findings
   (good move for conclusion, yes?)

FIGURE 8.12 General tips.

Titles

- Title your slide with a single, easily interpretable message

FIGURE 8.13 Titles.

Audience Analysis

- Who are they?
- Why are they here?
  - *key audience question: WIIFM?*
- What are their interests?
- What do they know?
- What do they want to know?
- Why do they want to know it?
  - *your key job: tell 'em! and tell 'em again!)*

FIGURE 8.14 Audience analysis.

Purpose

- Be clear about purpose:
  - informing
  - persuading
  - amusing

- Consider what you want audience to know, feel, or believe afterwards

FIGURE 8.15 Purpose.

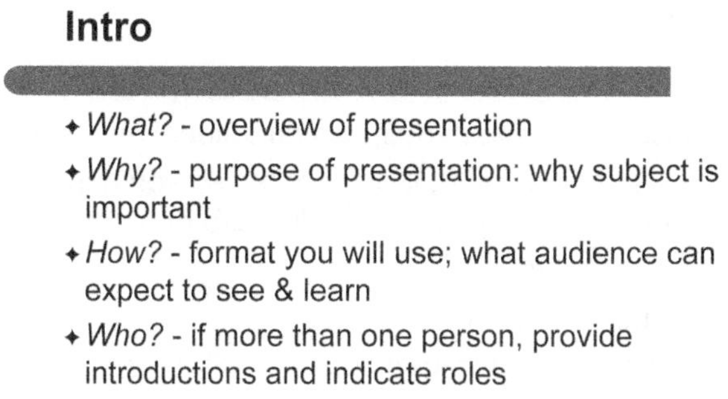

Intro

- *What?* - overview of presentation
- *Why?* - purpose of presentation: why subject is important
- *How?* - format you will use; what audience can expect to see & learn
- *Who?* - if more than one person, provide introductions and indicate roles

FIGURE 8.16 Introduction.

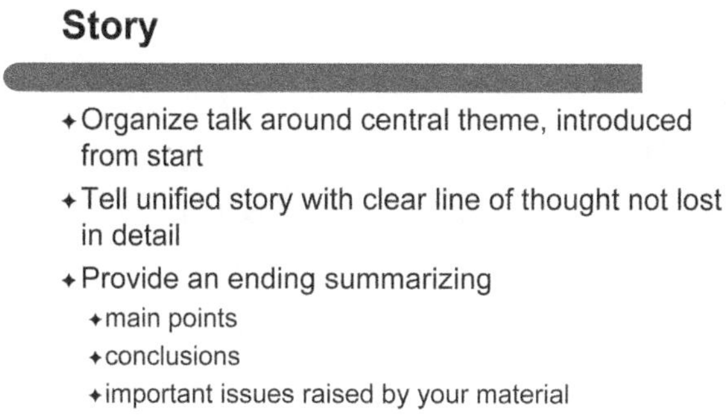

Story

- Organize talk around central theme, introduced from start
- Tell unified story with clear line of thought not lost in detail
- Provide an ending summarizing
  - main points
  - conclusions
  - important issues raised by your material

FIGURE 8.17 Story.

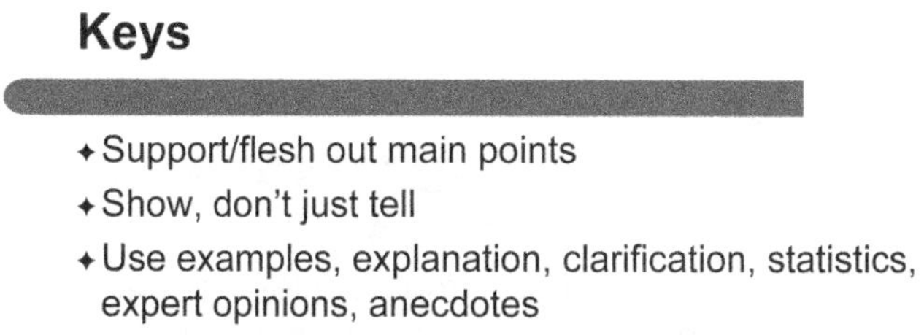

Keys

- Support/flesh out main points
- Show, don't just tell
- Use examples, explanation, clarification, statistics, expert opinions, anecdotes

FIGURE 8.18 Key points.

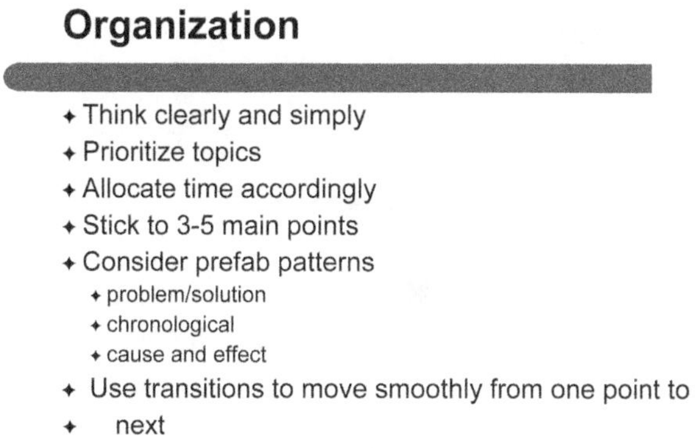

Organization

- Think clearly and simply
- Prioritize topics
- Allocate time accordingly
- Stick to 3-5 main points
- Consider prefab patterns
  - problem/solution
  - chronological
  - cause and effect
- Use transitions to move smoothly from one point to
- next

FIGURE 8.19 Organization.

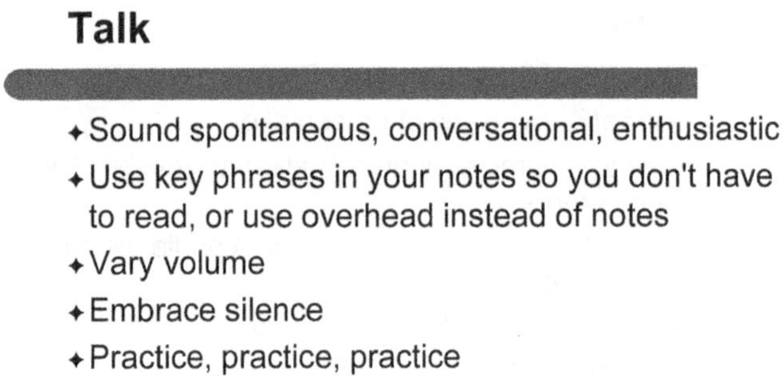

Talk

- Sound spontaneous, conversational, enthusiastic
- Use key phrases in your notes so you don't have to read, or use overhead instead of notes
- Vary volume
- Embrace silence
- Practice, practice, practice

FIGURE 8.20 Talk.

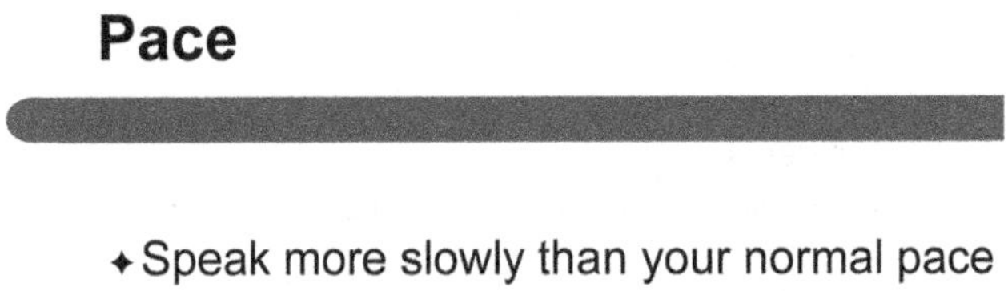

Pace

- Speak more slowly than your normal pace

FIGURE 8.21 Pace.

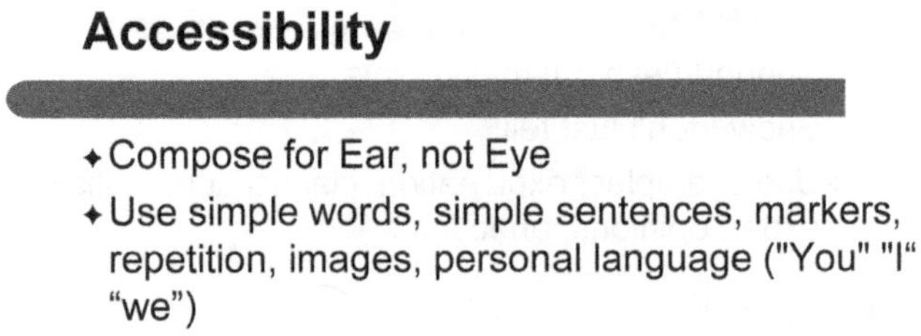

Accessibility

- Compose for Ear, not Eye
- Use simple words, simple sentences, markers, repetition, images, personal language ("You" "I" "we")

FIGURE 8.22 Accessibility.

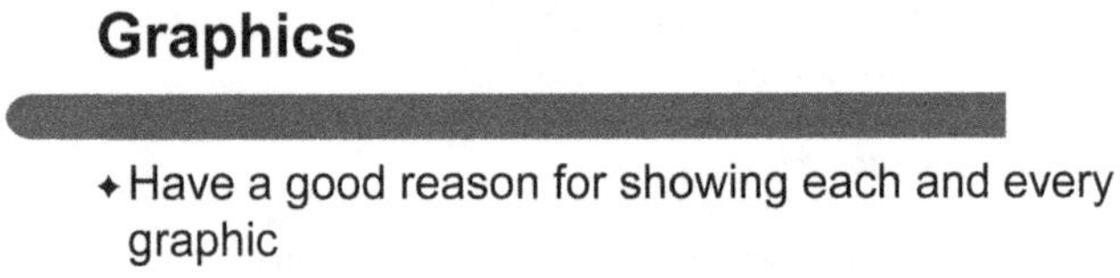

Graphics

- Have a good reason for showing each and every graphic

FIGURE 8.23 Graphics.

Fun

- Enjoy your presentation
  - "The person most committed to his or her attitude wins." (ok, it's from an infomercial....)
  - In other words, if you enjoy it, so may audience
- *Convey* enthusiasm (pleeeeeeeeease)

FIGURE 8.24 Fun.

Alert-ability

- Provide variety if possible
- Increase impact with novelty / uniqueness

FIGURE 8.25 Alertability.

Refocus

- Help us refocus in presentation even more than in written report:
  - "I will give the four basic reasons why change is needed."
  - "Now that we have analyzed the problem, we need to look at the possible solutions."
  - "The discussion so far leads to this final thought…"
  - "If you enact this program, two basic benefits will result…"

FIGURE 8.26 Refocus presentation.

Color

Beware of

- Certain combinations like red letters on blue background (causes "stereopsis")
- "Colorblind" combinations
  - Red/green
  - Blue/green

FIGURE 8.27 Color.

Color

- Contrast
  - Should be different enough, not yellow on white or brown on red
- Luminance/brightness
  - Balance: not bright blue and bright yellow (bright yellow on dark blue instead)

FIGURE 8.28 Color contrast and brightness.

Color

- Light on dark background in large rooms & with LCD projectors
- Dark on light background for overhead transparencies or small rooms
- 4 or less colors

FIGURE 8.29 Color choices.

Color

- Psychological effects
  - Red = anger or debt
  - Blue = calm, cold, relaxed
  - Warm colors (orange, red, yellow) can excite/stimulate

FIGURE 8.30 Color psychological effects.

Color

Colors most often confused:
- pink/gray
- orange/red
- white/green
- green/brown
- blue green/gray
- green/yellow
- brown/maroon
- beige/green
- pastels and muted tones

FIGURE 8.31 Confusing colors.

Conclusions, toughies

- "You have had much more time to work with your information than your audience; share your insight and understanding and tell them what you've concluded from your work"
- Review, highlight and emphasize - key points, benefits, recommendations
- Draw conclusions: where are we? ... what does all of this mean? ... what's the next step?

FIGURE 8.32 Conclusions.

Finally, please proofread and edit

- Spell-check!
- Look for inconsistencies:
  - Do you use end punctuation for some phrases and not others?
  - Do you capitalize first letters after a bullet for some items but not others?

FIGURE 8.33 Proofread and edit.

In short . . .

Present in way you'd like to be presented to

FIGURE 8.34 In summary.

PowerPoint

- Beware of doing something just because you can

FIGURE 8.35 PowerPoint misuse.

PowerPoint

- Can save as web page
- Can "pack & go"
- Can <u>link from one page to the web</u>
- Can link from one page in the presentation to another

FIGURE 8.36 PowerPoint features.

Thank you!

- Questions?

FIGURE 8.37 Questions and answers at the end.

Below are some suggested guidelines for presentations that can be customized for a specific capstone course:

- Establish a maximum amount of time (e.g., 15, 20, 25, or 30 minutes) for each presentation, including setup time, questions, and answers.
- It is strongly recommended to keep the duration of the presentation to 80% of the allotted time to allow sufficient time for setup (and everything that could go wrong) and Q&A period. Allow 1–2 minutes per slide.
- Make sure you invite your sponsor(s) to the presentations. Let your professor know that you have asked them and if they will attend.
- All members of the team must participate in the presentation. Assign slides to team members so they can practice their part and be ready to deliver a polished and professional presentation.
- Take notes at the end regarding the questions asked and comments and suggestions provided by your audience. Assign at least one member of your team to do this.
- Do not forget to have each team member of the design team introduce themselves and their role on the team at the beginning of the presentation.
- Include the following in your presentation:
  - Team organization and division of responsibilities
  - Definition of the problem and design specifications table
  - Design concept(s) and how it addresses the problem posed by the sponsor
  - Engineering analysis of the concept, testing, validation, and verification using mathematical modeling, relevant and modern engineering instruments, and software
  - Pugh and QFD analysis when appropriate

Documentation

- Trade-off analysis of alternatives
- Listing of problems and issues and identify which remain unresolved
- Explanation on how the design team plans to solve any unresolved issues
- Summary and conclusions
- The presentation should demonstrate creativity and innovation in the design solution and design team's accomplishments
- Provide a copy of your presentation electronic file (PowerPoint, etc.) to the capstone course professor and to the sponsor(s) before the scheduled time of your presentation
- The presentation will be rated on the effectiveness of the presentation of the project, including the team's ability as well as individual effectiveness to work as a team (among other factors)
- The team will be rated on the probability of achieving a successful design solution by the end of the capstone course
- The audience (other students, professors, mentors, and sponsors) should also participate by providing meaningful critique and comments for each team's presentation
- Presenters should remember that a positive attitude is essential in carrying out a successful design project and presenting the work

8.3 PHOTOS

Photographs have been used as sources of evidence and information in scientific endeavors for over a century. Over the past decade, the camera has become a ubiquitous device in every aspect of our lives. Every cell phone has a camera that is available to us at any instant to capture what is seen and observed. A photograph is a powerful tool to capture the progression of the design work, particularly for capstone design projects, but practically for all aspects of engineering. A picture can be more precise than many words written about a device or a design solution.

Photographs are persuasive because they capture the exact likeness of the things photographed. When we read about a design concept or how a process works, it is typically not entirely clear in our minds, and we have to rely on our internal mental envisionment of it. Not everyone reading the same text description will have the same mental picture. However, a photograph accurately captures an image that our brain can more readily and precisely interpret. Descriptive words and annotation on a photograph can enhance our understanding of the subject. A photograph is robust evidence of the design solution and how it was achieved.

In addition to photographing the stages of design evolution, assembly, parts, operation, maintenance, and fitting, people doing the work on the project should also be captured in the photograph. Photographing design team members in their roles working on the project documents their participation and contribution to the project. It serves as historical documentation of when the work was done, where, and by whom. People in photographs add value in understanding form, size, and function. Figure 8.38 shows an example photograph for a design project for an improved magnetic seal demagnetizer. The annotation in the figure helps in the understanding of the context and instruments used for testing the device.

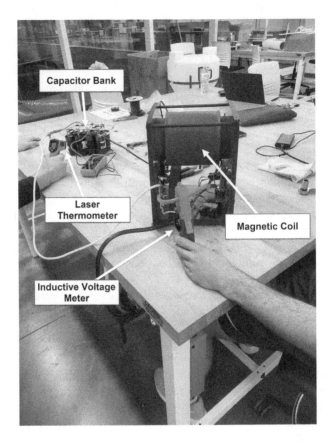

FIGURE 8.38 Example of annotated figure for magnetic seal demagnetizer project.

8.4 VIDEO PRESENTATION

The same cameras used for photography can typically be used for videography. Videography adds the dimension of time to the event being captured. The timeline captured in a video can be used to make calculations that are time-dependent such as a process timing study, measuring position, velocity, and acceleration.

A project video requires planning and setting goals for the session.

8.4.1 Timing Studies

Software tools are typically used for video time and motion analysis. Time and motion studies are used for process design analysis and improvement. A software tool accelerates the procedures for capturing the process and documenting the timing of events.

Using a software tool saves hours of tedious manual effort. The videography of the process must be coordinated with the process operators. The events, activities, or operational steps in the process must be documented to establish a baseline for comparison before and after design changes.

Video time and motion accurately document and time any task while simultaneously isolating the non-value-added work content. The video-supported analysis creates an unchallengeable history of the current process state.

8.5 POSTER

A poster presentation is a format that is commonly used at scientific and engineering conferences. Many engineering conferences have a special section for student presentations in a poster such as the ASEE zone or national conferences. It is a formal visual presentation of research or design. A design poster contains brief text, charts, graphs, and other visual aids to convey information about the project. The typical presentation format for a poster is where the attendees at the event view and read the poster while the design team is present and team members speak about the project and answer questions.

Technical posters can be of various sizes. The particular conference will specify the actual size. For ASEE, and many other student conferences, a size of 24 in. × 36 in. is recommended. The 24 × 36 size is a typical poster size, and it is easy to print to mid-range large format printers available at many engineering programs. Sometimes a 36-in. × 50-in. poster size is requested by a technical conference.

Posters are typically affixed to a corkboard or often mounted on a foam core poster board available in various sizes, including 24 × 36.

8.5.1 GUIDELINES FOR POSTERS

As part of the Capstone Design Course, you are required to prepare a technical presentation of your project in PowerPoint, a brochure/information sheet, and a poster form. This document provides some basic guidelines for the preparation of your posters.

Your poster must be prepared to present technical information about your project. A poster is simply a static, visual medium (usually of the paper and board variety) that you use to communicate ideas and information about your project. The material presented on your poster should convey the essence of your design project, including the problem that your design solution will solve, specifications, engineering analysis, technical drawings, illustrations, or solution photographs of your design, and your testing and operational results.

Before you start your poster project, stop and think!

8.5.1.1 How Much Poster Space Are You Allowed?

The purpose of poster presentations is not to have boards upon boards of information. You are restricted to a 24 inch × 36 inch poster size in portrait orientation. Make sure you include a small (12 point font) date on the bottom right of your poster. Your team can print your poster at the ECC after ultimately designing it and proofreading your material. The information presented on your poster must be accurate and without any errors.

8.5.1.2 Format

Make sure you include the following:

- Title, telling others the title of the project, the team members, and the sponsor.
- critical information such as date, sponsor(s), acknowledgments, appropriate URI logo, sponsor logo (ask permission)
- Summary of the design project stating your problem definition, design/solution approach, the critical features of your design, performance results, future enhancements, etc.
- An Introduction that should include clear statements about the problem, the design specifications, competitive solutions, and key features of your design. The introduction should lead to a description of project goals and objectives.
- Theory/Methodology/Approach section explains the basics of the engineering analysis technique you are using. You should also state and justify any assumptions so that your results can be viewed in the proper context.
- Design section that you use to show illustrative examples of the main designs created during your project.
- Testing section explaining how you tested your design. Did your design solution pass your tests?
- Conclusion section, listing the main findings of your project, and
- Further Work section that should contain your recommendations and thoughts about how the design could be improved, other tests that could be applied, etc.

Therefore, you have to present certain pieces of information but have limited space. So, before you rush away to the computer, spend a few moments or even hours planning your presentation.

8.5.1.3 Planning

Planning is essential. There are several stages in planning a presentation.

8.5.1.3.1 Gathering the Information

First, ask yourself the following questions.

- What are the objectives of the project?
- Has someone done the work before?
- How has your team approached the problem?
- Why did your team follow this particular approach?
- What are the principles governing the techniques your team is using?
- What assumptions did your team make, and what were your justifications?
- What problems did you encounter?
- What results did you obtain?
- Have you solved the problem?
- What did you discover?
- Are your analyses sound?

Although the above list is by no means exhaustive, it gives you a rough idea. You have to stand back and think again about the Whats, the Hows, and the Whys of the project that you are doing. You have to critically examine the approach you have taken and the results you have obtained.

Ideally, it would be best to do this throughout your design project. In doing so, you will have a clearer idea of the objectives and contributions you have, or have not, been able to make.

It would be best if you brainstormed within your team. Write your answers on a large piece of paper, not necessarily in an ordered fashion. The intention is to note as many points as possible not to miss any important aspects. The ordering and pruning of the information come later. From your list, note the common areas, topics, or pieces of information, and group them. Use color or number coding or circles and lines to help you identify and categorize the information. This activity should help you focus further on the content you can use confidently.

8.5.1.4 Content

If you follow the above guidelines, then the content is more or less determined by you. However, given that you have limited space, you now have to decide between what is essential and what is not. Your decision should be based on at least two factors, namely:

- What are you trying to achieve by presenting the posters? Is it to sell a design? Is it to tell people what your team has done? Is it to say to people of a new invention or discovery? Is it to convince people that your design is better than another or solves a particular problem?
- Who will be attending the presentation? Are they technical people? What is the level of their knowledge of your design project?

The answers to these questions define the type of content to include and set the presentation's tone.

8.5.1.5 Design

An advertising billboard is a poster. If well designed, it will be attractive and engender a lasting impression; earnest but not dull. Importantly, it should shout out to you – "buy me!" or you would think, "I want that!" Similarly, in using posters to convey technical information, they should be designed such that readers think "Yes!" or "I see!" and leave with the impression that they have learned something new.

Some rules of thumb in poster design:

1. Plan
2. Keep the information simple
    - make full use of the space, but do not cramp a page full of information, as a result, it can often appear messy
    - be concise. Use only pertinent information to convey your message
    - be selective when showing results. Present only those that illustrate the main features of your project. However, do keep other results handy so that you may refer to them when asked
3. Use colors sparingly and with taste
    - Colors should be used only to emphasize, differentiate, and add interest. Do not use colors to impress!

- Try to avoid using large swathes of bright garish colors like bright green, pink, orange, or lilac.
- Pastel shades convey feelings of serenity and calm, while dark bright colors conjure images of conflict and disharmony.
- Choose background and foreground color combinations with high contrast and complement each other - black or dark blue on white or very light grey is good.
- It is better to keep the background light as people are used to it (for example, newspapers and books)
- Avoid the use of gradient fills. They may look great on a computer display, but the paper version can look poor unless you have access to a high-resolution printer.

4. Do not use more than two font types
   - too many font types distract, especially when they appear on the same sentence
   - fonts that are easy on the eyes are Times-Roman and Arial.
   This is Times-Roman
   This is Arial
5. Titles and headings should appear larger than other text but not too large. The text should also be legible from a distance, say from 4 ft to 6 ft.
6. Do not use all UPPER CASE types in your posters. It can make the material difficult to read. Just compare the two sentences below:
   EXAMPLE OF A LINE WHERE ALL THE CHARACTERS ARE IN UPPER CASE.
   Example of a line where only the first character of the first word is in the upper case.
7. Do not use a different font type to highlight essential points. Otherwise, the fluency and flow of your sentence can appear disrupted.
   - Use underlined text, boldface, or italics or combinations to emphasize words and phrases.
   - If you use bold italicized print for emphasis, then underlining is unnecessary.
8. Equations
   - should be kept to a minimum
   - present only the essential equations
   - should be large enough
   - should be accompanied by terminology to explain the significance of each variable
9. A picture is worth a thousand words ... (but only if it is adequately drawn and used appropriately)
   - graphs
   - choose graphs types that are appropriate to the information that you want to display
   - annotations should be large enough, and the lines of line graphs should be thick enough so that they may be viewed from a distance
   - do not attempt to have more than six line graphs on a single plot

- instead of using lines of different thicknesses, use contrasting colored lines or different line styles to distinguish between different lines in multi-line graphs.
- multi-line plots or plots with more than one variable should have a legend relating the plotted variable to the color or style of the line.
- diagrams and drawings,
- should be labeled
- annotate pictures to explain what you are showing
- drawings and labels should be large and clear enough so that they are still legible from a distance
- do not try to cramp labeling to fit into components of a drawing or diagram. Use "arrows" and "callouts" for annotations.
- clipart should only be used if it adds interest to the display and complements the subject matter. Otherwise, they distract attention from the focus of the presentation.
- can also be "dangerous" as you may spend more time fiddling about with images and choosing appropriate cartoons than concentrating on the content.

10. Check your spelling!
11. Maintain a consistent style
    - inconsistent styles give the impression of disharmony and can interrupt the fluency and flow of your messages.
    - headings on the different pages of the poster should appear in the same position on all pages.
    - graphs should be of the same size and scale, especially if they are to be compared.
    - if bold lettering is used to emphasize one page, then do not use italics on others.
    - captions for graphs and drawings should be positioned below the figure
    - titles for tables should be positioned at the top
12. Arrangement of poster components should appear smooth
    - you are using a poster to tell a story about what you have done and achieved. The way you arrange the sections should follow the design project timeline.
13. Review
    - make draft versions of your poster sections and check them for
    - mistakes
    - legibility and
    - style
    - try different layout arrangements (for example, it may not be best to put the conclusions on the bottom right where it may be difficult to read)
    - ask your teammates, friends, colleagues, or professors for their "honest" opinions
    - be critical

8.6 TECHNICAL INFORMATION SHEET

Compressing details about the design project into a single page information sheet is an efficient and professional way for quickly conveying information about the project. The information sheet is a professional way to communicate and is typically used for professional meetings and conferences.

The technical information sheet (TIS) should provide essential facts about the product or process designed in a compressed document that makes it easy and quick for technical professionals to understand the problem and the design solution. Information sheets play an essential role in communicating information with all stakeholders of a project. It is a convenient way to make sure stakeholders understand the critical parts of the design solution quickly and effectively.

An information sheet is a one-page document, typically double-sided, consisting of information and data about the product or process design. An information sheet lists all the critical information, facts, and figures on the design solution, visually, with charts, drawings, schematics, and images.

The project information sheet should contain design problem definition, design solution information, statistics, design specifications, and technical data. The information sheet should be created with clear, crisp, and concise information. The TIS is laid out in a visual format to emphasize key points and achievements.

Before creating an information sheet, clear guidelines and purpose should be created by the professor for capstone.

8.6.1 Guidelines for Capstone Technical Information Sheets

There are a few things one must keep in mind before creating a capstone TIS:

- It should preferably be a single page long and double-sided. This format is a suitable format for the professional engineering context.
- The information presented in the CTIS should be brief and concise.
- It should visually appeal and draw attention, comprising tables, charts, drawings, schematics, graphs, and bullet points.
- It should be composed of a balance of visuals, words, graphics, and illustrations.

The CTIS is used for some essential purposes:

It Saves Time – The CTIS is typically one double-sided page. Therefore, it helps save much time for the interested stakeholder by providing precise and concise information on a single page. In addition, because CTIS are designed and created to be visual, the reader can quickly skim through and comprehend information, data, and facts quickly and efficiently.

It Is Easy to Read – The textual information in a CTIS should comprise a balance of white spaces, bullet points, and bold headings. It is more of a visual document, so essential information is presented using tables, drawings, graphs, pictures, schematics, and charts that are easy to read and understand.

8.6.2 Style Guide for Capstone Technical Information Sheet

A style guide should be developed for the CTIS for each capstone design based on what is needed. The following sections provide some of the suggestions:

Project Title – The CTIS title should be the project title. University Logo and sponsoring company logs can be added for project affiliation.

Design Problem Description and Motivation – The problem definition and motivation for solving the problem is an essential section of the information sheet. It should be brief and precise, setting the stage for the remaining information in the document.

Design Specifications – The design specifications should be listed in a concise quantitative tabular form.

Product or Process Design Process – The document should briefly describe the process and steps taken by the design team to achieve a design solution, including any alternative and promising design solution concepts considered.

Design Solution Achieved, Tested, and Redesigned – The document should briefly describe the design solution achieved. Summary of testing results for proving the design solution should be included. The final design solution, which is the result of redesign cycles and testing, should be presented.

Conclusions and Recommendations for the Future – The significant features and results from the design project are presented in the conclusion section. Recommendations for future work guide stakeholders and especially the sponsor regarding what can be done to improve on the design in future developments.

8.7 PRELIMINARY DESIGN REPORT

The PDR is a formal engineering document that encapsulates the entire accomplishments of the design team during the first phase of the design project, typically at the end of the first semester of the project. The PDR documents the design team's work, covering the design team's entire work, leading to the creation of proof concept design and a prototype.

The PDR is a complex document with many aspects that must be included, so a set of guidelines is necessary to prepare the document for each project. The typical length of a PDR is 50–150 pages. The following section covers a list of suggested guidelines for creating the PDR.

8.7.1 Guidelines for Preliminary Design Report

General Format – PDF format is an excellent form for PDRs in a single file. It can be shared easily through file-sharing or e-mailed. It is also a standard format that maintains file consistency across various operating systems and printing systems. Most sponsors appreciate receiving the PDR in this format. A 1-in. margin (on all sides) on each page is recommended. Include a page number on every page. Outline the report in the manner prescribed below.

Title Page – Project title, team number, team name, team logo, team members (and role of each member), sponsor, name of a faculty advisor(s), academic year, submission date, and a report identification number (if assigned). No page number should be inserted on this page.

Abstract – The abstract is a concise and brief description of the project. It should include project objectives, what was accomplished, and how in the design process. Typically, the abstract should not exceed 500 words and should be limited to one page. Page # "ii."

Table of Contents – Contains a list of major and minor headings within the report along with page numbers. Page # "iii."

List of Acronyms – Include a list of all acronyms used in the report. Page # "iii."

List of Tables – Include a list of all tables in the report, including the page number of each table. Each table must be numbered. Page # "iii."

List of Figures – Include a list of all figures in the report and the page number of each figure. Each figure must be numbered. Page # "iii."

Introduction – Include project description and problem definition, the stated design requirements and expectations, a brief chronology of previous work performed in this area, the purpose, scope, and objectives of the design work. Page # "1".

Literature Searches – Provide a complete list of literature researched by each team member and referenced in the PDR. Explain what was used or learned from each article or publication. Explain how each publication or report pertains to the team's design solution.

Patent Searches – Provide a complete list of patents researched and used as a reference in the design project. Patents listed should be relevant to the design project. Explain how each pertains to the design solution.

Project Planning – Describe the project planning and management. Include a Gantt chart showing the timeline, tasks, milestones, and resources showing team member task assignments. Specify tasks completed as a percentage, milestones, and managerial tools implemented to achieve design goals

Design Specifications – Describe how customer requirements and engineering criteria were transcribed into engineering design specifications. Describe how target values were set for the engineering design specifications. The target values for design specifications should be numerical parameters with units. In addition to any write-up, set the design specification in a tabular format.

Conceptual Design – Describe the process the team followed to generate design solution concepts. Include a list (including sketches) of all concepts generated by each team member. Analyze each concept and evaluate its applicability as a solution to the design problem. Finally, perform a Pugh analysis on all concepts and include the decision matrix to explain how it was created. Show the iterations on the Pugh analysis to select three or four top design choices.

Competitive Analysis – Document the market analysis conducted to identify and assess the competition (or alternatives to the team's top design) and establish a strategy to gain market/customer advantage. Present a QFD analysis of the design solution as a tool to identify important aspects of the project. Include a trade-off and competitive analysis in the detailed discussion of the QFD analysis.

Project Specific Details and Analysis – This section includes any details, data collection activity, or engineering analysis specific to the team's project. For example, product design teams should include market analysis, demand forecasting, cost versus price information, and surveys of potential users. Process design teams should include a flowchart or floor plans of the current process and time studies of the current operations. National design competition teams should document evidence of work done to fulfill the requirements of the competition that goes beyond typical capstone course requirements.

Product or Process Design Details – Describe how the chosen concept was developed into the prototype design. Include drawings, including composing parts. Drawings may be placed in the Appendices if too numerous but should be referenced in this section. Organize the appendix material so it is clear what is included. Include dimensions, tolerances, annotations, linkages, and relationships.

Engineering Analysis – Describe mathematical models such as process models, thermal, fluids, structural, vibrational, static, dynamic, materials, and finite element analysis, as appropriate. Is the product design sound, safe, and following the current state of knowledge (state of the art), regulatory requirements, standards, and codes?

Proof of Concept – Discuss the proof of concept, prototype, or how the design was proven that the design functions as envisioned. Compare the prototype's features, functions, and performance against all of the design specifications the team created.

Financial Analysis – Describe the analysis of design solution costs for implementation as a usable product. Include person-hours (and dollar equivalent) spent on the project by team members, faculty advisors, consultants, etc. Include an analysis of costs for mass production of design, market demand, forecast of technology, and future revisions. Analyze return on investment as appropriate. Discuss cost savings and manufacturing or process efficiencies as a result of the project.

Conclusions – Describe how the design satisfies the design specifications.

Further Work – Describe what is yet to be accomplished as a continuation or follow-up on the project. Describe the project plan for continuation and further development. Describe the project conclusion goals.

References – List all publications, papers, and reports, in a reference list.

Appendices – Include important information that is too bulky to fit within the main report, such as computer programs and detailed assembly drawings. All items in the appendices must be referenced in the main text.

Format the report following a professional engineering organization publications guideline such as the example suggestions listed below (source: ASME publication standards with modifications):

Footnotes

- Numbered consecutively using superscript numbers
- Positioned flush left at the bottom of the column/page in which the first reference appears
- Footnote text should be 10 pt.
- Spacing: allow one extra line between the text and the footnote

Equations

- Insert equations apart from the body of the text and center. Use two or three-line spaces to separate equations from the text.

- Number equations consecutively, using Arabic numerals enclosed in parentheses and position flush right along the final baseline of the equation.

Graphics

- Graphics include photographs, graphs, and line drawings
- Number consecutively and include a caption
- Include a centered caption using 11 pt. boldface serif typeface, uppercase, below the graphic
- Annotations within the graphic should be no smaller than 9 pt.
- Use enough space to separate graphic from the text and other elements
- Position the graphic as close as possible to where it is referenced within the body of the report
- Size the graphics as follows:
  - 7.5 in. across the top of the page
  - 9×7.5 in. (length $\times$ width) to fit the entire page (portrait or landscape)
- Gantt and QFD charts should be formatted as large as possible (tabloid is best 11×17 in.)

Tables

- Number consecutively and add a descriptive caption
- Include 11-pt. boldface caption centered above the table
- Position the table as close as possible to where it is referenced within the body of the report
- Use enough space to separate the table from text and other elements
- Size the tables as large as needed to display all information clearly
  - 7.5 in. to fit across the top of the page
  - 9×7.5 in. (length $\times$ width) to fit the entire page
  - Rotate the table sideways (landscape) if needed for a better format

List of References

References to sources for cited material should be listed together at the end of the report under a heading called References; footnotes should not be used for this purpose. References should be arranged in numerical order according to their order of appearance within the text.

1. Reference to journal articles and papers in serial publications should include:
   - last name of each author followed by their initials
   - year of publication
   - full title of the cited article in quotes, title capitalization
   - full name of the publication in which it appears
   - volume number (if any) in boldface (do not include the abbreviation, "Vol.")
   - issue number (if any) in parentheses (do not include the abbreviation, "No.")
   - inclusive page numbers of the cited article (include "pp.")

2. Reference to textbooks and monographs should include:
   - last name of each author followed by their initials
   - year of publication
   - full title of the publication in italics
   - publisher
   - city of publication
   - inclusive page numbers of the work being cited (include "pp.")
   - chapter number (if any) at the end of the citation following the abbreviation, "Chap."
3. Reference to individual conference papers, papers in compiled conference proceedings, or any other collection of works by numerous authors should include:
   - last name of each author followed by their initials
   - year of publication
   - full title of the cited paper in quotes, title capitalization
   - individual paper number (if any)
   - full title of the publication in italics
   - initials followed by last name of editors (if any), followed by the abbreviation, "eds."
   - publisher
   - city of publication
   - volume number (if any) in boldface if a single number, include "Vol." if part of a larger identifier (e.g., "PVP-Vol. 254")
   - inclusive page numbers of the work being cited (include "pp.")
4. Reference to theses and technical reports should include:
   - last name of each author followed by their initials
   - year of publication
   - full title in quotes, title capitalization
   - report number (if any)
   - publisher or institution name, city

Sample References

1. Ning, X., and Lovell, M. R., 2002, "On the Sliding Friction Characteristics of Unidirectional Continuous FRP Composites," ASME J. Tribol., 124(1), pp. 5–13.
2. Barnes, M., 2001, "Stresses in Solenoids," J. Appl. Phys., 48(5), pp. 2000–2008.
3. Jones, J., 2000, Contact Mechanics, Cambridge University Press, Cambridge, UK, Chap. 6.
4. Lee, Y., Korpela, S. A., and Horne, R. N., 1982, "Structure of Multi-Cellular Natural Convection in a Tall Vertical Annulus," Proc. 7th International Heat Transfer Conference, U. Grigul et al., eds., Hemisphere, Washington, DC, 2, pp. 221–226.
5. Hashish, M., 2000, "600 MPa Waterjet Technology Development," High-Pressure Technology, PVP-Vol. 406, pp. 135–140.
6. Watson, D. W., 1997, "Thermodynamic Analysis," ASME Paper No. 97-GT-288.

7. Tung, C. Y., 1982, "Evaporative Heat Transfer in the Contact Line of a Mixture," Ph.D. thesis, Rensselaer Polytechnic Institute, Troy, NY.
8. Kwon, O. K., and Pletcher, R. H., 1981, "Prediction of the Incompressible Flow Over A Rearward-Facing Step," Technical Report No. HTL-26, CFD-4, Iowa State Univ., Ames, IA.
9. Smith, R., 2002, "Conformal Lubricated Contact of Cylindrical Surfaces Involved in a Non-Steady Motion," Ph.D. thesis, http://www.cas.phys.unm.edu/rsmith/homepage.html.

8.8 FINAL DESIGN REPORT

The FDR is a formal engineering document that is the culmination of the complete accomplishments of the design team during the entire design project at the end of the capstone design course(s). The FDR captures the entire design project from the beginning problem definition to creating and validating a design solution.

The FDR is a comprehensive document that includes much of the content documented in the PDR. Therefore, a set of guidelines is necessary to prepare the document for each project. The FDRs are typically 100–400 pages in length. The following section covers a set of suggested guidelines for creating the FDR.

8.8.1 Guidelines for the Final Design Report

General Format – Prepare the FDR in pdf format as a single file. Allow a 1-in. margin (on all sides) on each page. Include a page number on every page. Outline the report in the manner prescribed below. The pdf file is an efficient format to share with the project stakeholders, including the sponsor.

Title Page – Include project title, team number, team name, team logo, team members (and role of each member), company sponsor, name of the faculty advisor(s), submission date, and a report identification number. No page number should be inserted on this page.

Abstract – Write a concise and informative summary description of the project, its objectives, what was accomplished, and how. This section should not exceed 500 words and should be limited to one page.

Table of Contents – List major and minor headings within the report along with page numbers.

List of Acronyms – Include a list of all acronyms used in the report.

List of Tables – Include a list of all tables in the report, including the page number of each table. Number each table sequentially.

List of Figures – Include a list of all figures in the report, including the page number of each figure. Number each figure sequentially.

Introduction – Include project description and problem definition, the stated design requirements, and expectations, a brief chronology of previous design work performed, the purpose, scope, and objectives of the design project.

Patent Searches – Provide a complete list of all patent searches. List only patents that are relevant to the project. Include patent drawings where appropriate.

Evaluation of the Competition – Describe the market analysis to identify the competition (or alternatives to the final design) and how a strategy was established to gain sponsor approval or market advantage.

Engineering Design Specifications – Describe how the customer requirements and engineering criteria were transcribed into engineering design specifications. Describe how target values for design specifications were established. List design specifications numerically, including units.

Conceptual Design – Describe the process that was used to generate concepts for design solutions. Then, list and illustrate all concepts that were generated as viable solutions to the design problem. Next, describe the analysis and evaluation of each concept. Finally, describe the Pugh analysis of the concepts and cycles of Pugh analysis to select top design solution choices.

Quality Function Deployment – Describe the QFD analysis of the project as a tool to identify important aspects of the design project. Include a trade-off and competitive analysis in the detailed discussion of the QFD analysis. Show how the QFD was updated and evolved throughout the design process.

Design for X – Explain how the design was achieved under the constraints of the Xs such as safety, cost, manufacturability, reliability, ergonomics, and the environment.

Project Specific Details and Analysis – Include project-specific details in this section, including data collection activity, engineering analysis, market analysis for a product, sponsor survey for process design specific to the team's project. For example

- Product design teams should include market analysis, demand forecasting, cost versus price information, and surveys of potential users.
- Process design teams include flowcharts, process diagrams, floor plans of the current process, and time studies of the current operations.
- National student design competition teams should document work done to fulfill the design competition requirements that go beyond typical course requirements.

Product Design Details – Describe how the chosen concept was developed into the final design. Include drawings and bill-of-materials for product design projects. Include process charts and a description of all of the steps and flow.

Engineering Analysis – Explain and describe in detail all forms of engineering analysis, for example, process efficiency, thermal, fluids, structural, vibrational, static, dynamic, materials, and finite element analysis, as appropriate. Explain how the product design is sound, safe, and following the current state of knowledge (state of the art), regulatory requirements, standards, and codes.

Build, Manufacture, Create – Present the steps taken in creating the design solution, which can be a product or process. For product design, describe the manufacturing analysis detailing the most efficient manufacturing and assembly course. For process design, describe the details of the process that was designed. Finally, describe how each can be scaled to a typical production level.

Testing – Describe how the design was verified and validated. Include the detailed test matrix for all components and subsystems of the overall design solution. Explain any system-level tests performed. Show categories for tests performed, e.g., performance tests, design specification compliance, safety, life expectancy, failure modes, disassembly, decommissioning, and disposal. Demonstrate and explain the application of engineering standards for product compliance (e.g., ASTM, FCC, OSHA)?

Redesign – Describe the redesign cycles of the design solution based on what was learned from the testing procedures. Explain the rationale for redesign and how it was optimized. Explain the testing procedures and how the results were applied for redesign and optimization. Explain the evolution of the redesign process into the final design. Present the final design achieved from the redesign process. If time constraints did not allow implementing the redesign plan, i.e., rebuilding the prototype, detail the steps necessary to achieve an improved redesign. Provide detailed recommendations for future design teams to improve on the design based on what was learned.

Project Planning – Describe the project planning and management. Include a Gantt chart showing the timeline, tasks, milestones, people, and resources, including team member task assignments. Specify tasks completed as percentages, milestones, and managerial tools implemented to achieve design goals

Financial Analysis – Provide a detailed description of projected and actual costs for the project. Include sources of funding. Include person-hours (and dollar equivalent) spent on the project by team members, advisors, mentors, consultants, and sponsors. Include an analysis of costs for scaling up the bench-scale design solution to a usable product or process, include market demand, forecast technological evolution, and future revisions. Analyze return on investment as appropriate.

Operation/Usage – Describe the operation of the product or usage of the process design. Guide a prospective user of the design solution through all operation steps and procedures, similar to an operator's manual or a user's guide. Explain safety procedures and guidelines. Write a safety guide for users if appropriate. Did you produce an operator's manual? Did you produce a safety guide/manual; an assembly manual; a repair manual?

Maintenance – How will your product be maintained and/or serviced? How will it be disposed of after reaching the end of its useful life?

Additional Considerations – Perform a critical assessment of broader aspects of your project. If there is no additional impact, explicitly, state so for each of the five categories listed below, and provide tangible evidence that there is no impact. This section MUST include:

- Economic impact
- Societal impact
- Ethical considerations
- Health, ergonomics, and safety considerations
- Environmental impact and sustainability considerations

Conclusions – Document that your design does indeed satisfy the design specifications. State all significant findings, design rules, potential for commercialization of your product, and next steps.

References – List of all papers, reports, etc. referenced in the text.

Appendices – Include important information that are too bulky to fit within the main report, such as computer programs, detailed or assembly drawings. Everything in the appendices must have been referred to in the main text.

Format your report in accordance with ASME publications guidelines listed below (with minor modifications):

Footnotes

- Numbered consecutively using superscript numbers
- Positioned flush left at the bottom of the column/page, in which the first reference appears
- Footnote text should be 10 pt.
- Spacing: One extra line between the text and the footnote

Equations

- Display equations should be set apart from the body of the text and centered. Use two or three line spaces to separate equations from the text.
- Numbered consecutively, using Arabic numerals enclosed in parentheses and positioned flush right along the final baseline of the equation.
- No ellipses (dots) from the equation to the equation number, or any punctuation at the end of the equation itself.

Graphics

- Includes photographs, graphs, and/or line drawings
- Numbered consecutively and captioned
- Caption = 11 pt. boldface serif typeface, uppercase, centered below the graphic
- Callouts within the graphic should be no smaller than 9 pt.
- Spacing: use enough space to separate graphic from text and other elements
- Positioning: within body of the report after first reference
- Sizing: graphics should be sized for the final publication
  - 7 1/2 in. – to fit across the top of the page
  - 9 × 7 1/2 in. (L × W) – to fit the entire page
  - 6 1/2 in. × 8 1/2 in. (L × W) – across the entire page sideways (turned)
- Gantt chart and QFD chart are exempt from this rule

Tables

- Numbered consecutively and captioned
- Caption = 11 pt. boldface serif typeface, uppercase, centered above the table
- Callouts within the table should be no smaller than 9 pt.
- Positioning: within body of the report after first reference
- Spacing: use enough space to separate table from the text and other elements
- Sizing: tables should be sized for the final publication

- 7 1/2 in. – to fit across the top of the page
- 9 × 7 1/2 in. (L × W) – to fit the entire page
- 6 1/2 in. × 8 1/2 in. (L × W) – across the entire page sideways (turned)

Text Citation – Within the text, references should be cited in a numerical order according to their order of appearance. The numbered reference citation should be enclosed in brackets.

Example

It was shown by Prusa [1] that the width of the plume decreases under these conditions.
- In the case of two citations, the numbers should be separated by a comma [1,2]. In the case of more than two reference citations, the numbers should be separated by a dash [5–7].

List of References – References to original sources for the cited material should be listed together at the end of the report under a heading called References; footnotes should not be used for this purpose. References should be arranged in a numerical order according to their order of appearance within the text.

1. Reference to journal articles and papers in serial publications should include:
    - last name of each author followed by their initials
    - year of publication
    - full title of the cited article in quotes, title capitalization
    - full name of the publication in which it appears
    - volume number (if any) in boldface (do not include the abbreviation, "Vol.")
    - issue number (if any) in parentheses (do not include the abbreviation, "No.")
    - inclusive page numbers of the cited article (include "pp.")
2. Reference to textbooks and monographs should include:
    - last name of each author followed by their initials
    - year of publication
    - full title of the publication in italics
    - publisher
    - city of publication
    - inclusive page numbers of the work being cited (include "pp.")
    - chapter number (if any) at the end of the citation following the abbreviation, "Chap."
3. Reference to individual conference papers, papers in compiled conference proceedings, or any other collection of works by numerous authors should include:
    - last name of each author followed by their initials
    - year of publication
    - full title of the cited paper in quotes, title capitalization
    - individual paper number (if any)

- full title of the publication in italics
- initials followed by last name of editors (if any), followed by the abbreviation, "eds."
- publisher
- city of publication
- volume number (if any) in boldface if a single number, include, "Vol." if part of larger identifier (e.g., "PVP-Vol. 254")
- inclusive page numbers of the work being cited (include "pp.")

4. Reference to theses and technical reports should include:
   - last name of each author followed by their initials
   - year of publication
   - full title in quotes, title capitalization
   - report number (if any)
   - publisher or institution name, city

Sample References

1. Ning, X., and Lovell, M. R., 2002, "On the Sliding Friction Characteristics of Unidirectional Continuous FRP Composites," ASME J. Tribol., 124(1), pp. 5–13.
2. Barnes, M., 2001, "Stresses in Solenoids," J. Appl. Phys., 48(5), pp. 2000–2008.
3. Jones, J., 2000, Contact Mechanics, Cambridge University Press, Cambridge, UK, Chap. 6.
4. Lee, Y., Korpela, S. A., and Horne, R. N., 1982, "Structure of Multi-Cellular Natural Convection in a Tall Vertical Annulus," Proc. 7th International Heat Transfer Conference, U. Grigul et al., eds., Hemisphere, Washington, DC, 2, pp. 221–226.
5. Hashish, M., 2000, "600 MPa Waterjet Technology Development," High Pressure Technology, PVP-Vol. 406, pp. 135–140.
6. Watson, D. W., 1997, "Thermodynamic Analysis," ASME Paper No. 97-GT-288.
7. Tung, C. Y., 1982, "Evaporative Heat Transfer in the Contact Line of a Mixture," Ph.D. thesis, Rensselaer Polytechnic Institute, Troy, NY.
8. Kwon, O. K., and Pletcher, R. H., 1981, "Prediction of the Incompressible Flow Over A Rearward-Facing Step," Technical Report No. HTL-26, CFD-4, Iowa State Univ., Ames, IA.
9. Smith, R., 2002, "Conformal Lubricated Contact of Cylindrical Surfaces Involved in a Non-Steady Motion," Ph.D. thesis, http://www.cas.phys.unm.edu/rsmith/homepage.html.

Part II

Build, Test, Redesign, Repeat, Document

9 Building from Proof of Concept Design

The proof of concept design is typically substantiated with a prototype to validate some aspects of the overall design achieved. The next step is typically to advance the development of the design by creating a better working model of the design concept. This working model may be a physical build in the case of a product, or it may be a process model implemented in a modeling and simulation software, e.g., ProModel.

The working model will be based on the prototype for proof of concept. In some cases, it is a natural progression of the evolution of the proof of concept prototype into a more mature build that is functional and usable. In other cases, it is a fresh start based on what was learned from building the prototype. Regardless, it is essential to develop plans, tasks, and milestones for creating the working model.

9.1 LOGISTICS OF BUILDING THE DESIGN

Building the design concept beyond the prototype requires planning and resources. The resources include people, time, materials, software or instruments, and facilities. Arranging for the resources needed requires planning and having a fully developed design plan. Before a build process can start, the sponsor and the capstone professor should review the proposed build plan and review it with the team. Any anticipated problems should be discussed during the review process and resolved to the stakeholders' satisfaction (the team, sponsor, and capstone professor).

The design team should decide the scope of the build to match the requirements and available resources. It is always a good plan not to consume all of the available resources on the first build because, typically, the first designs have many aspects that need to be corrected later in the process. Additionally, design iterations always result in improvements.

The timeline is strictly constrained in capstone design projects, so it is essential to plan the build project steps. The milestone for completing the working model is typically one month into the second semester of capstone. This deadline is necessary to allow time for testing and redesign. The one-month period for building the design is highly challenging for design teams. Many activities must be completed, such as creating a build-plan, purchasing materials, scheduling labs and resources, and assembly. The purchasing step must be carefully overseen because it involves submitting purchase requests, receiving approval, ordering, shipping, and receiving. When the team has a plan and list of materials, they should shop the materials to find the best possible prices for any items they need.

The team will have a budget, but the goal is to plan not just for the initial build of the working model but also for all the expenses (and time) associated with testing and redesign.

A build-space is also part of the planning. Suppose the project is a software development project. In that case, space is in the form of computer or cloud storage needs, other software for developing the model, and perhaps some physical space (such as a computer lab or meeting room) where the team can do their group work. For a process design project, the project team needs to coordinate with the sponsor to access the process to collect data or experiment with the design concepts. When process changes are being designed for an active manufacturing line, it may not be possible for the team to test their design ideas on the actual line so the team could develop a software model of the process or alternatively create a mock process to test ideas.

For product development projects, the physical space needed will depend on the size of the working model. Some projects are small, and others may need some floor space. The larger the physical size of the model, the more coordination and planning are needed. Product prototypes typically need other resources such as a machine shop, electronics lab, 3D printing, welding, measurement, and testing equipment. Access to those resources and facilities may be limited because they are shared resources, and many design teams need to access them. Design teams should schedule their use of shared facilities and resources as soon as possible to prevent scheduling conflicts so that their project can proceed smoothly.

9.2 RESOURCES NEEDED

Once a plan for the build of the design has been created and reviewed by stakeholders, the design team must identify all resources needed to complete the task. Resources needed depend on what the team plans to construct, software, a process, or a product prototype.

For a software development project, the team may need the following:

- Computer systems for the development of the software
- A choice of an operating system for the design platform
- A software development environment including a compiler/interpreter and libraries
- A user interface system that is available through open-source software or specific to a vendor such as Windows, Mac OS, IOS, Android, Unix, and Chrome

The school or the sponsor typically provides the computer systems and software development environment. Students may also decide to use their personal hardware and software for development.

A target process is selected for process design projects for new development or possible modification of existing software. The process may be based on a physical system. For example, the sponsor may be a hospital asking for process improvement to admit new patients into an emergency room facility. Resources that may be needed for a process design project may include:

Building from Proof of Concept Design 153

- Access to the review and document the current process to collect data and create a dynamic model of the process
- If the process location is at the sponsor facility, then transportation to and from the facility is one of the resources that the team needs to manage
- Access to people involved in the operation of the current process is needed to conduct interviews and learn about the details of the process and what is expected from design changes
- Software for simulation such as ProModel, MATLAB, Visual Components, MapleSIM, and MATLAB
- Video capture hardware such as a camera, although cell phones work great for capturing process videos
- Computers capable of running the simulation and modeling software efficiently

For product development projects, a critical decision is scaling the working model of the design. Should the working model be full-scale subject to loads, pressures, and temperatures of the actual device? Or should the model be a smaller-scale representation of the design with some extreme conditions removed to enable the team to build a working model? Typical design projects with industry sponsorship can fall in either category of full-size build or scaled-down version. The design team must coordinate the scope of the working model with the sponsor and other stakeholders for the project.

Cost and project budget is a significant driver for deciding to scale down the design-build and eliminate some features that are considered to be less critical in testing the primary function and form of the design working prototype. The amount of time needed to build is another major factor in the scope of the building phase of the design project. Time is a critical component that in capstone design is a constrained and fixed period.

Resources needed for a product development project include:

- Materials in raw form in basic shapes such as rectangular, sheet, rod, tube, and square as needed for the design, which means the design must use these elemental shapes
- Tools that are available to process the materials such as saw, drill, sander, milling machine, lathe, welder, and soldering equipment
- Fasteners such as screws, nuts and bolts, clamps, and pins
- Adhesives and sealants appropriate for the design and environmental conditions
- D printing and materials for the 3D printer such as PLA, TPU, PC, PLA+, and ABS

Additionally, all types of projects require the following resources:

- Lab space for group work on the design project
- Meeting space to meet with the sponsors and stakeholders

9.3 ACTUAL VERSUS MODEL

Capstone projects often involve complex real-world problems with many interdependencies with interfacing systems. The design project may encompass the actual product or process or a more simplified, isolated, and smaller-scale implementation that fits the budget and schedule of the capstone program.

9.4 MATERIALS

Choice of materials plays a fundamental role in the product or device design process. Many different facets must be considered such as safety, strength, service requirements, fabrication requirements, cost, and chemical compatibility. The designers can ensure safety for humans and the environment by selecting sustainable and safe materials that can perform during the product's life and after the useful life when the product can be recycled and materials reused.

Figure 9.1 shows the relationship between materials, form, function, and material performance.

Product designers create a list of materials called a bill of materials (BOM) to create the design and build plans for the working model. The BOM includes a detailed listing of all materials that are used in the design. Some examples of materials include:

- Wood and wood products: pine, balsa, oak, and MDF
- Mesh fabrics (e.g., Kevlar, carbon fiber mesh)

FIGURE 9.1 Relationship of materials to process, form, and function.

Building from Proof of Concept Design

- Metals and Alloys: Aluminum alloys, Mild steel, Alloy steels, Stainless steels, Cast iron, Copper, Brasses, Nickel alloys, Titanium alloys, Magnesium alloys, Zinc alloys, Lead
- Polymers: Nylon, PET, HDPE, PVC, Polypropylene, POM, Acetal, ABS, PLA, Polythene, Polystyrene, Polycarbonate, Porcelain
- Ceramics: Alumina, Brick, Concrete, Silicon, Silicon carbide, Diamond, Titanium Carbide, Zirconia
- Composites: carbon fiber epoxy, fiberglass epoxy, mixed carbon fiber Kevlar composite
- Coatings, epoxy, and paints
- Adhesives, glues, silicone, waterproof adhesives

The designers create a list of materials for each part as well as for the entire design. Each part and the assembly should include a list of the type of material such as wood, plastic, metal, composite, fabric, glass, and foam. Each type of material can be assessed for safety and sustainability. In addition to safety and sustainability, materials must be analyzed for chemical compatibility, strength, flexibility, hardness, and thermal characteristics. Each material is characterized by the manufacturer with material safety data sheets (MSDS). The MSDS for the entire product should be created and included with the design package and documentation.

When the design team is working with the materials for their design, they must collect and document the MSDS and quantities of the materials stored in the lab for emergencies and fire safety. In case of a fire, the firefighters need access to the list of materials stored in the capstone lab to use appropriate fire retardants for the chemistry present. The U.S. Occupational Safety and Health Administration (OSHA) requires MSDS for all products. The MSDS can be used to assess the chemical hazards of the materials used in the design and proper methods of protecting the designers while building with those materials.

9.4.1 Materials Selection Charts

A quick and powerful way to identify materials for design projects is a graphical way represented in material selection charts. Material mechanical characteristics extend an extensive range of values, so typically, log-log charts are used to represent the numerical values of the data. 2D charts are used to compare the different dimensions of the application of materials. Some different ways showing material property parametric relationship include:

- Young's modulus versus density
- Young's modulus versus cost
- Material strength versus density
- Material strength versus toughness
- Material strength versus strain
- Material strength versus cost
- Material strength versus maximum or minimum temperature
- Specific stiffness versus specific strength
- Electrical resistivity versus cost

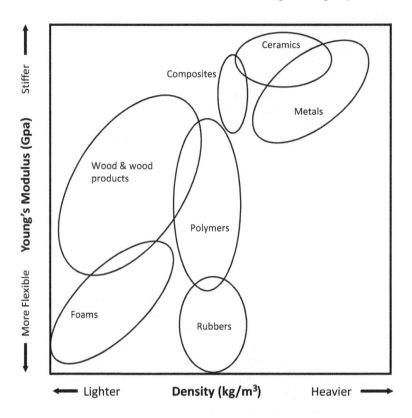

FIGURE 9.2 Material selection chart – Young's modulus versus density.

Figure 9.2 shows the comparison of Young's modulus for different classes of materials in the form of a bubble chart. The chart shown does not have a scale but shows the relative position of each material class. A search through online sources will find more detailed charts that can be applied to particular design needs. Young's modulus and density are material properties related to the atomic arrangements and density within the material. Young's modulus depends on the atomic bonds and arrangement in the material. Flexibility and stiffness are measures of how much a material deforms when a load is applied. Young's modulus has a strong correlation with density (note the diagonal orientation of the bubbles on the chart). The chart scales is logarithmic in both horizontal and vertical directions.

Composites offer high stiffness and are lighter compared with metals and ceramics. Some applications, such as structural members in a building, require stiff materials. Light materials that are also stiff are rare but very useful for applications where weight is a primary consideration such as an aircraft or a quadcopter drone.

Figure 9.3 shows a chart of Young's modulus versus cost. This chart indicates that some metals are less expensive than other materials and offer high stiffness.

Building from Proof of Concept Design

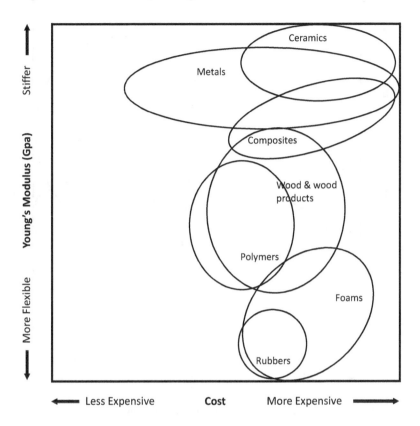

FIGURE 9.3 Material selection chart – Young's modulus versus cost.

Figure 9.4 shows material strength versus toughness. Ductile materials are less brittle (tougher) and controlled by plastic deformation. Ceramics are strong in compression (weaker under tension) but brittle.

Strength is a measure of the material not failing under some applied stress. The strength in the chart is measured for materials under tension except for ceramics which are measured under compression load. Toughness is proportional to the energy required to propagate a crack through the material. Toughness is a descriptor for materials under an impact force. Material composition and processing significantly affect the strength and toughness of the material, particularly for metals. Metals typically deform under an impact force instead of being brittle, so they are considered tough. Cast iron is typically brittle because of the high carbon content formed in flakes during the casting process. Carbon steel becomes hard but brittle during a quenching process. Quenching and tempering carbon steel increase strength and toughness, making it suitable for applications such as hammers, engine components, and saw blades.

Plastics and polymers are considered to have a medium to high toughness, but their performance is sensitive to cracks and defects. Plastics and polymers are ductile but are good choices for many products because they absorb energy under load or impact before failure (Figures 9.5–9.8).

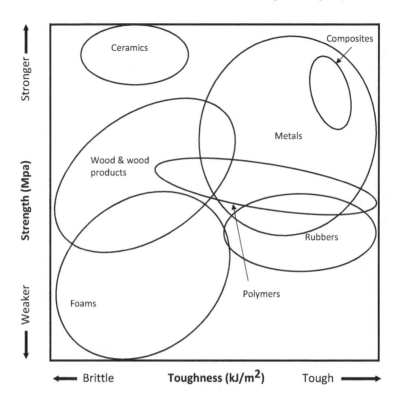

FIGURE 9.4 Material selection chart – strength versus toughness.

9.4.2 Metals

The service requirements may lead to the choice of metals for manufacturing the working model of the design. Metals are typically solid at room temperature (Mercury being an exception). Metals have a reflective surface and can be polished. Metals are ductile and malleable so they can be formed into different shapes such as sheets, rods, tubes, and blocks. Metals are much harder materials than many alternatives and are typically good conductors of heat and electricity. Metals have high density and are heavy. The melting and boiling points of metals are high, and therefore, they are suitable for high-temperature applications.

If corrosion is a consideration, then certain metals such as aluminum or stainless steel are good choices. Aluminum is lighter and softer than stainless steel and has a lower melting point.

Some common metals of interest include:

- Hot-rolled steel
  - High-strength plate
  - Angle, channel, beam, flat bar, square bar, square tube, round bar, rectangular tube, pipe, sheet, plate, wire, rebar, circles, rings, and gussets are some of the forms that steel material can be purchased
  - Galvanized steel is corrosion resistant

Building from Proof of Concept Design

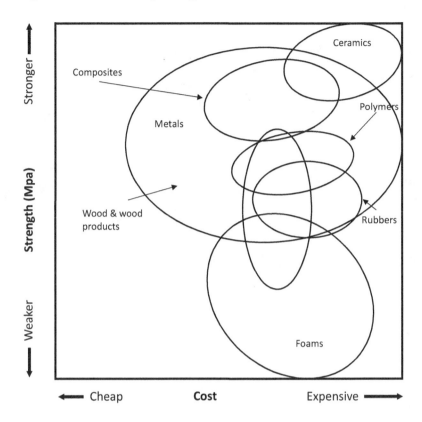

FIGURE 9.5 Material selection chart – strength versus cost.

- Cold-finished steel
  - Alloys of interest include 1018 and 1045
  - Flat, round, square, hexagon, shaft, and keyed shaft are some of the forms that can be purchased
- Tool steel
  - High hardness steel
  - Rod and flat are the forms that can be purchased
  - Used for creating blades and drill bits
- Aluminum
  - Aluminum alloys of interest include 6061, 6063, 3003, 5052
  - Angle, channel, beam, flat bar, square bar, round bar, square tube, round tube, pipe, sheet, and plate are some of the forms available
  - Aluminum can be anodized to add corrosion resistance and color to the surface finish
- Stainless steel
  - Stainless steel alloys of interest include 304, 315, 303, and 304
  - Angle, channel, flat, square, round, shaft, keyed shaft, square tube, rectangular tube, round tube, pipe, sheet, and plate are some of the forms available
  - Stainless can be polished to a highly reflective surface

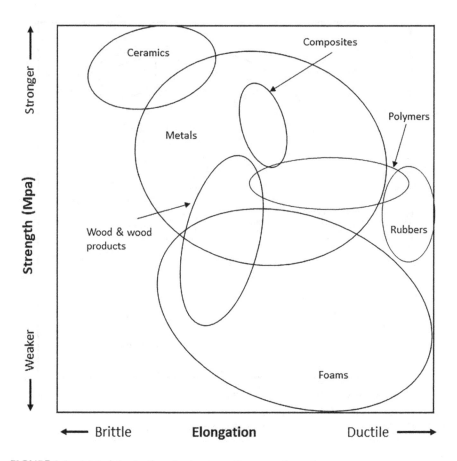

FIGURE 9.6 Material selection chart – strength versus elongation.

- Brass
  - Brass alloys of interest are 360 and 260
  - Flat bar, square bar, round bar, sheet, and plate are some of the forms available
- Copper
  - Copper has very high thermal and electrical conductivity
  - Flat bar, square bar, round bar, sheet, and plate are some of the forms available

9.4.3 Plastics

Plastics have an extensive range of applications in engineering and in creating working models for capstone design. Plastics are easy to cut, drill, glue, form, and weld. The forms that are available to purchase include sheet, rod, tube, and various profiles. Some plastics are also used for 3D printing materials such as ABS, PLA, polycarbonate, PETG, and PEEK. Plastic materials used in capstone design include the following:

Building from Proof of Concept Design

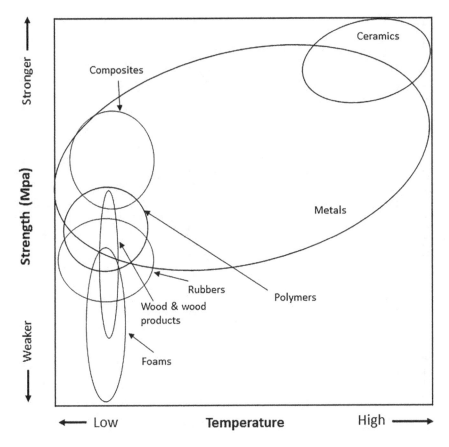

FIGURE 9.7 Material selection chart – strength versus maximum temperature.

- ABS
- Acetal
- Acrylic
- ColorCore
- Delrin
- ETCFE
- Flex tubing
- FRP panels
- HDPE
- Kydex
- LDPE
- Noryl
- Nylon
- PEEK
- PETG
- PFA-M
- Phenolic

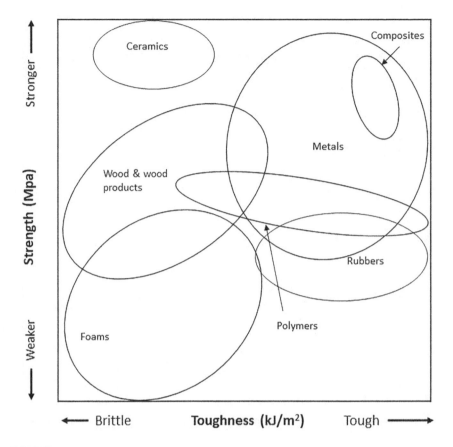

FIGURE 9.8 Material selection chart – strength versus toughness.

- Polycarbonate
- Polypropylene
- PPS
- PTFE
- PVC
- Styrene
- Turcite
- UHMW
- Ultem

The choice of plastics to use depends on the form, application, hardness, flexibility, strength, thermal properties, and cost.

9.5 MANUFACTURABILITY

Manufacturability is a measure of how effectively a product can be manufactured. Design, application, time, cost, technology, machine shop, 3D printing, and skill

Building from Proof of Concept Design

sets are critical parameters in manufacturability. The degree of manufacturability is a direct result of how the product was designed. Therefore, a product should be designed for manufacture, reducing cost and optimizing product fitness for the design specifications.

Many different methods of manufacturing can be applied to a designed product. Design team members have different skill sets, so part of the analysis of manufacturability is to assess how those skills apply to different methods of manufacturing a working model of the design.

9.5.1 Manufacturing Process Selection

Choice of materials, working model scale or size, availability of machine tools, cost, and skills drive the selection of a particular method for manufacture. For example, if the team decides to build their prototype from steel, weight, size, welding, fasteners, machining, finishing, and cost are some of the issues to consider. Comparing manufacturing alternatives is necessary to understand the manufacturing issues to select an optimal method for the team's design. Analysis of the manufacturing includes steps to outline and model the entire manufacturing process and estimate time, cost, materials, and requirements for each scenario. A project plan for the manufacture is an excellent way to capture all of the steps involved and calculate total time and cost before jumping into the manufacturing step. Creating a project plan for the manufacture forces the design team to think through all of the steps and create scenarios for different approaches to building the design.

9.5.2 Design for Manufacturing and Assembly (DFMA)

Design for manufacturing (DFM) is a design process where the product or process design is modified and optimized for making the product manufacturing faster, better, and cheaper. The design process is examined, refined, and optimized at every step in DFM analysis. DFM analysis includes optimizing the manufacturing process, design, materials, and testing and verification aspects. Each design project is unique and requires a specific analysis of the manufacturing issues for that project. In capstone design, typically, one working model is created, so the DFM analysis needs to be conducted before manufacturing the working model.

DFM analysis consists of a set of design principles, each aiming to improve the product, reduce costs, and reduce the manufacturing time.

Parts reduction is a direct way to improve the product, reduce costs, and reduce assembly time. Fewer parts mean fewer items that have to be purchased, which directly reduces the overhead associated with the procurement processes, handling, and storage. Fewer parts reduce the complexity of all aspects of the product during its life cycle. If two parts do not move relative to each other, then they can be combined into a single part. Combining parts eliminates the need to assemble and fasten them and can reduce costs. For example, a washer and a bolt could be combined into one part (a flanged bolt), eliminating the need to install the washer.

Modular design is another way to simplify and reduce costs. Modular design can reduce costs by either using off-the-shelf modules or modules from past or similar design activities. Manufacturing, assembly, testing, redesign, purchasing, repair, operation, and service can be simplified by using a modular design. Testing and redesign can be simplified because the modules can be independently tested when using standard modules.

Standard parts, modules, and components are robust and reliable because they have already been designed and tested. Standard components can be purchased, and therefore, their use reduces total design and development time for the entire project. Standard parts also cost less compared to developing them from scratch. Using standard off-the-shelf parts and components reduces variability in the overall product design and improves total quality.

Standard off-the-shelf components and parts may be selected in such a way that they serve multiple purposes in the overall design. The use of multipurpose parts or components can reduce the part count. An example of a multipurpose component is replacing a 3D-printed wall that is a part of the enclosure for battery and electronics, driving a brushless motor with aluminum to serve as a structural wall, and dissipating heat from the enclosure by conduction. Multipurpose components can simplify the design and increase functionality.

Standard parts and components can help design teams improve quality, reduce costs, and achieve a design faster. Standard parts can often be used in multiple design projects. Designers develop expertise and experience with features and performance of specific parts they have used before, accelerating their design work on new projects. Manufacturing costs are reduced by purchasing parts in bulk or large quantities. An inventory of frequently used parts also reduces the overhead costs and times associated with ordering and receiving operations. A capstone design lab staff can identify commonly used parts by design teams and make purchases at the beginning of the academic year to save on costs and reduce prototyping time for the design teams. For example, many teams use an Arduino or a RaspberryPi microprocessor for their projects. These can be made available to the teams that need them early.

Manufacturing and fabrications costs (and time) can be optimized during the design phase. Finishing operations such as polishing and painting are time-consuming and costly and can be minimized. Tolerancing is a common problem. Tight tolerancing significantly increases machining and part production costs. Loose tolerancing may reduce the quality of the product and adversely affect its function. Design teams should carefully examine the tolerancing needs for their design project and allow time for iterations to find an optimal value.

Fasteners increase the number of steps, time, and therefore the cost of assembly. Fasteners are typically installed manually by workers and increase handling time. Installing fasteners may also require special tools, for example, a riveting tool, which may need to be purchased and maintained. Tabbing and alignment features may need to be designed into each part to improve assembly and fastening precision. Fasteners can sometimes be replaced with snap fits or tabs. Each fastener should be carefully examined to eliminate it by redesigning the part and the assembly method. If fasteners have to be used, the variety should be minimized to reduce time and minimize errors in assembly and fastening.

Building from Proof of Concept Design

During the design phase, each part's fabrication and manufacturing aspects should be considered to make fabrication and assembly straightforward such that errors can be mitigated. Suppose the fabrication and assembly are too complex and require special skills or training. In that case, the costs can quickly escalate and significantly increase the manufacturing of the working model before the testing phase can begin. To the extent possible, the use of outside services for fabrication and manufacturing should be reduced to keep cost and timeline in check.

9.5.3 Mistake Proofing

Mistakes can be costly to a product or process especially when they result in failure. Mistakes can happen during the design phase, during assembly or execution, or by the end-user of the product or process. Mistake proofing or poka-yoke (Japanese equivalent) is any method that prevents an error from occurring or makes it obvious when the mistake has occurred. Mistakes occur because of human involvement in a process. When an error happens, it may result in an apparent failure which may create a dangerous situation and cause harm to people or assets. A cause-and-effect analysis can help the designers identify the root cause of the failure caused by the error.

Mistake proofing starts with an analysis of the process by creating a process flowchart. Each step in the process is reviewed to identify the potential for human errors. During the analysis, it is essential to think outside the box and expand the thinking to all possible ways the process user may take an unanticipated action and cause a malfunction or failure. For each possible error discovered through testing and analysis, create a feature in the design to prevent that error. If it is not possible or too costly to change the design to prevent the error, create user instruction or training to mitigate user error in operation. Alternatively, create user instructions or training to self-check their work and prevent the error. Automation may be used in some cases to bypass the error-prone step.

An example of mistake-proofing is the data cable connectors for computers and cell phones. The universal serial bus (USB) connectors type A was designed with a block tab to allow the connector to be inserted into a USB plug only in one orientation to assure the correct electrical connection for data and power. The USB A design is an example of a poka-yoke design. The user can only insert the USB plug only in one orientation. This would prevent failure if the user were allowed to insert the USB plug the wrong way incorrectly. A plastic block tab prevents the error. However, commonly, the user attempts to put the USB plug in one way, and 50% of the time, they cannot insert the USB plug because of the tab. They have to flip the USB plug and insert it the correct way. An improved design of the USB-A plug connector was introduced as USB-C. The electrical contacts are replicated on both sides in USB-C, so it does not matter if the user flips the connector. The new design is a poka-yoke design, and it improves on the previous USB-A design by eliminating the 50% user failure to insert the plug (Figures 9.9–9.12).

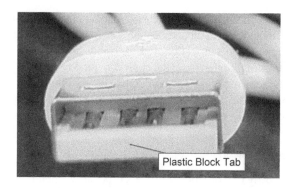

FIGURE 9.9 Poka-yoke example USB-A connector.

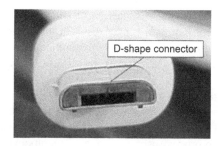

FIGURE 9.10 Poka-yoke example USB-micro connector.

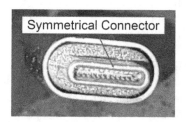

FIGURE 9.11 Poka-yoke example USB-C connector.

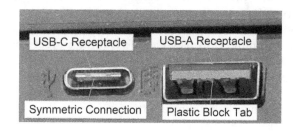

FIGURE 9.12 Poka-yoke example USB-C connector.

9.6 3D PRINTING

3D printing has become popular over the past decade because of critical technologies that have reduced prices and increased accuracy and usability. 3D printing has become an effective way to prototype and create working models of many different varieties of designs. The design created in the form of CAD must be specially created to allow for efficient printing and optimize the use of materials and printable geometries. Some geometries cannot be printed on a 3D printer, such as when the walls are too thin or print features that result in internal geometries that are too small. Printing usable 3D parts also require knowledge of the materials and best practices to achieve a successful and usable print.

Objects are created layer by layer on a 3D printer. Standard lower-priced printers extrude a melted plastic filament sintering to the top of the previous layer of plastic. 3D printing is also called additive manufacturing instead of machining, where the material is removed (subtractive manufacturing). The print head in a 3D printer typically moves in the x-y plane, and the glass build plate moves in the z-direction.

The fused deposition modeling (FDM) technology was patented in 1989 by S. Scott Crump and Lisa Crump. They later co-founded Stratasys, Ltd. Their technology was based on feeding a plastic filament into a heated extruder layering the material by moving in the x-y direction. The fundamental patents expired in 2005, opening the field for many other companies manufacturing FDM printers.

Most standard 3D printers use fused filament deposition manufacturing (FFDM) technology. The filament is typically 1.75 mm (or 2.85 mm), with the print-head nozzle being 0.4 mm. The user's layer height can be controlled in the slicing software but should not exceed half the nozzle diameter. Most printers print in a Cartesian coordinate system. Some printers are non-Cartesian.

3D printers use a microcontroller to control the movement of the print heads and the print bed. An Arduino controller is sufficient as the microcontroller because the computational power needed for motion calculations is relatively light. The microcontroller software is called firmware. A standard open-source firmware is called Marlin.

The process flow for 3D printing is shown in Figure 9.13. The designed part in CAD must be exported in a portable file format known as stereolithography (STL). The slicing software choice depends on the type of printer. Simplify3D is a manufacturer of independent software that supports many different types of printers. Many printer vendors provide slicer software specific to their printers free of charge. Check with the staff responsible for the 3D printers at your school or organization for the needed slicer software. The slicer software can read the STL file and require the user to enter parameters to customize the print for the specific printer and materials used for the print.

Standard 3D printer slicing software requires the model file to be in stereolithography (STL) format. The STL format is old and inefficient, but it is broadly accepted as the norm. The CAD software can export the design model in several different formats, including STL. The STL file is an approximate representation of the detailed geometry in the CAD model. The approximation is achieved by converting the model surfaces to a triangle mesh and outputting the triangle vertices coordinates and the normal vector to the triangle plane to the STL file in 3D Cartesian coordinates. Therefore, the STL files are large and compute-intensive to process. The user can set

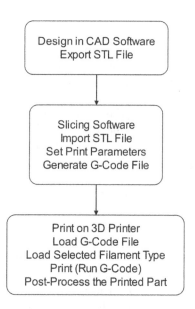

FIGURE 9.13 3D printing process.

a system of units and precision intervals to adjust the resolution and size of the STL file. For 3D printers, it is best to use the highest resolution model allowed by the CAD software (Figure 9.14).

Because of converting a precise geometry into an STL approximation, issues may arise in the STL file, such as small triangles, super long triangles, the surface not being watertight, and incorrectly oriented surface normals. Most slicing software includes options to repair the model or repair the mesh. If the mesh is problematic beyond the capabilities of the slicing software, another software program may be used to repair the mesh. Meshlab (www.meshlab.net) is an example of a software program that can manipulate, repair, and optimize meshes in an STL file. Simplify3D is another example of software that is efficient in mesh repair and slicing STL files for many different types of printers (Figure 9.15).

3D printers need software instruction sets, called gcode, to execute a print. The slicer software creates gcode for the specific printer from the STL model file and user input for printer and print parameters. The slicing software needs user input for printer characteristics such as the print bed dimensions and z travel distance, nozzle size installed, and temperatures for the nozzle and the print bed (Figure 9.16).

We discuss the usage and settings using Simplify3D as a typical slicer for 3D printing. Other slicers have similar features, but the details of menus and wording options are unique to each software. Simplify3D has the ability to write the output g-code file to an SD memory card which can then be loaded onto the 3D printer. Another option is to connect the computer running Simplify3D to the 3D printer using a USB cable. If connected by the USB cable, the computer running Simplify3D software controls the 3Dprinter for set up and motion for each step.

Under the tools menu, the machine control panel is used to set up the serial port for communication to display and control all aspects of the 3D printer (Figure 9.17).

Building from Proof of Concept Design

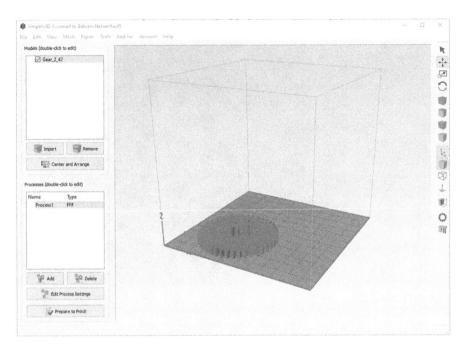

FIGURE 9.14 STL file loaded in the Simplify3D software.

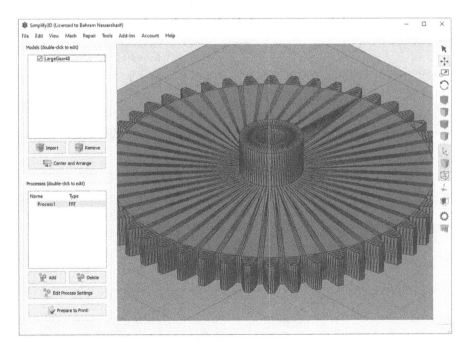

FIGURE 9.15 STL file mesh view (line drawing) in the Simplify3D software.

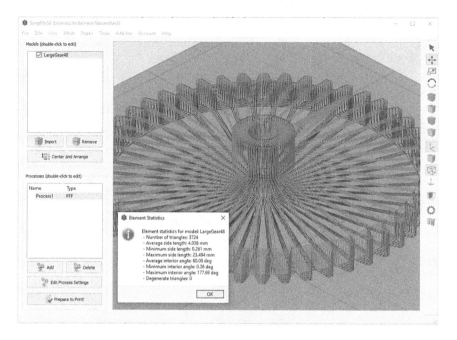

FIGURE 9.16 STL file statistics in the Simplify3D software.

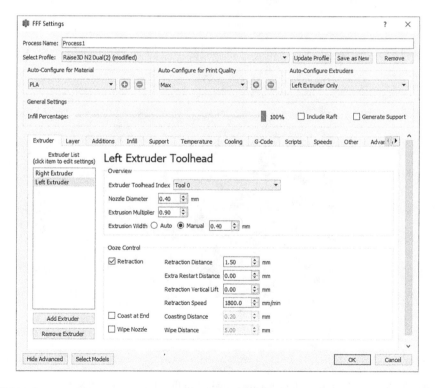

FIGURE 9.17 Process settings (Extruder) in Simplify3D software.

Building from Proof of Concept Design

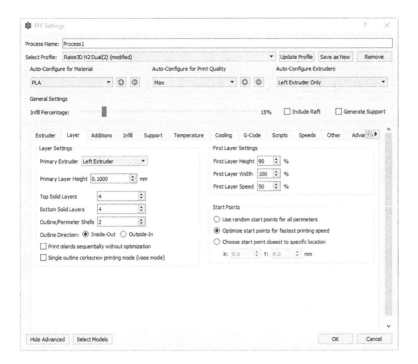

FIGURE 9.18 Process settings (layer height) in Simplify3D software.

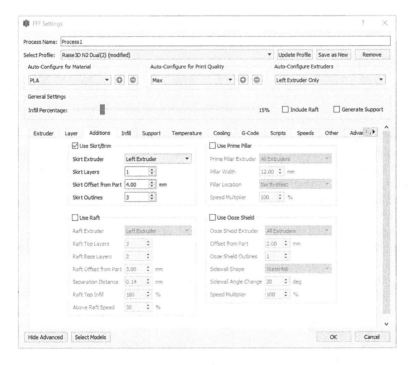

FIGURE 9.19 Process settings (Additions) in Simplify3D software.

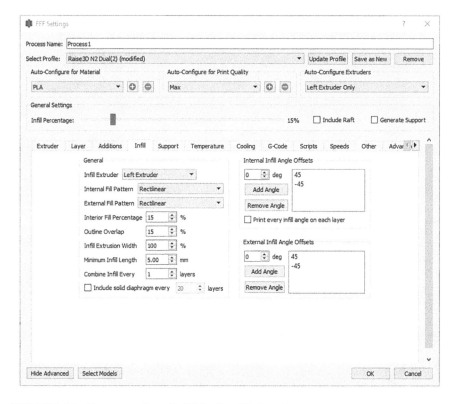

FIGURE 9.20 Process settings (Infill) in Simplify3D software.

9.7 QUALITY

Engineering students' quality of a design-build of a product or process is greatly dependent on their skill sets gained from outside the academic program. In general, engineering programs do not train their students to be builders or process engineers ready for a production-ready creation of their designs. Some students may have gained such experience through co-op or internship experiences. Those students will be a great asset to the student design team because not only can they engage in a more complete creation of their design product or process, but they can also teach their teammates to engage more deeply in the experience.

When the team engages in the design process, they should also account for their experience and skills to build their design product or process. If the team's skill set is not matched with their final design, then they may not be able to prove their design concept or build a working solution of usable quality.

Suppose the desired level of implementation quality cannot be achieved with the design solution that the team has produced. In that case, the team must make changes in the redesign stage to use appropriate methods or tools to create a working version of their solution.

Assessment of the quality of the working model should be made by the project sponsor, professor of the capstone course, and team mentors and advisors. Keeping

Building from Proof of Concept Design 173

the judgment on quality outside of the team allows for a more objective evaluation of the quality of the team's implementation of their design ideas.

It is essential that the engineering students try to make/implement their design solutions and not have someone else build/implement their designs. Engineering students need to deeply understand the implications of their design choices with respect to manufacturing or implementation. Suppose the students are allowed to outsource their design solutions. In that case, they will not learn from all of the changes that must be made during the manufacturing or implementation of their design solution. Learning from the changes is an essential part of the design process.

9.8 BILL OF MATERIALS

The engineering BOM is created by the engineering design team to detail the product as designed. BOM consists of the list of assemblies, sub-assemblies, parts, components, and materials necessary to build the design. BOM can be output from CAD software such as SolidWorks if the CAD contains the necessary details, or the design team may manually create it. The BOM must be complete such that anyone can build the product if they acquire the items on the list and have a set of instructions to assemble the design. The BOM should list all necessary items in a manner that those items can be acquired or procured from common sources.

A detailed BOM allows the product manufacturers to prepare for the manufacturing steps necessary to create the product and mass produce it. Special part and assembly requirements can be incorporated into the manufacturing process if the BOM is complete and detailed. The product and manufacturing management team can find the best parts pricing, quality, and quantity to shorten supply-chain lead times and mitigate any problems or delays during product manufacturing. The management can also engage in the competitive procurement of items needed for product manufacture based on a detailed and concise BOM.

The completeness and accuracy of the BOM enable the manufacturing group to minimize glitches and delays in product manufacturing. Manufacturing rework can become very expensive because of either parts errors or misinterpretation of information in the BOM.

The completeness and accuracy of BOM in the capstone design project reflect the maturity and completeness of the design work performed by the student design team. As such, reviewing the BOM in the design documentation is a strong indicator of the quality and completeness of the design achieved.

The sample design reports that are part of the electronic complements to this book contain many good examples of BOM tables.

9.9 PROCUREMENT

Items listed in the BOM have to be procured at some point. The student design teams must purchase the items for their design project, prototyping, building the working model, testing, and redesign. Each university or school has its own set of policies for purchasing. State universities tend to be more bureaucratic in purchasing policies.

Many schools issue and allow the use of a purchasing credit card (P-card). A P-card can greatly expedite purchasing small parts and items, which is typical of many student design project needs.

A typical project may have expenditure needs as low as $300.00 and ranging into several thousand dollars depending on the project's scope. It is an excellent idea to have an internal process for the capstone course to manage the procurement needs. There can be an enormous amount of clerical and paperwork needed to process and document the purchasing needed for capstone design for a large class. Teams should keep track of their purchase requests and alert and remind the people responsible for reviewing, approving, and processing purchase requests by the team to obtain the materials they need on time. This administrative aspect of capstone design is necessary and an excellent experience for young engineers as they prepare for what comes next in a typical engineering job when they have to account for and report things to their management.

10 Redesign

Redesign is a natural phase for all engineering design projects. After building a working model and testing it, a list of fixes and problems is typically generated. What is learned from the testing of the design model will guide the design team in correcting any errors, problems, and making improvements to the design.

Design is an iterative process. Redesign is the iteration phase where problems and errors can be fixed, and the design solution can be optimized or improved. The planning for redesign will include many considerations, including:

- Review the design specifications for any changes that may be appropriate based on what was learned from building the working model and testing.
- Review the test results for specific items to be addressed.
- Review any sponsor/customer interaction experience results for guiding the inclusion of specific features requested by the sponsor after testing.
- Consider user training as part of a possible solution to the design.
- Review budget and resources for adequacy to scope the redesign activity.
- Identify unresolved issues and develop solutions to resolve those issues.
- Coordinate with the sponsor to scope out the project redesign.
- Update the project plan to include the scope of the redesign.

The redesign process never ends for real-world design projects. The designers continuously identify items to fix and optimize. However, the design team must satisfy project deadlines and budget constraints. The iteration on the design should be guided by first solving the critical problems and then optimizing the solution.

The team should capture and document the sponsor/customer requirements and specific design features that helped the sponsor achieve their objectives. What specific features of the design helped the sponsor or user achieve success? Should those aspects become a more significant part of the design? Were those design aspects apparent to the user? Consider enhancing those valuable aspects of the design. The improvements should be tested again, and user responses must be collected to check for improvements and enhanced satisfaction with the design solution. Was the user experience better? Did they use the design solution more efficiently and effectively?

The design team should also capture and document the sponsor/customer experiences that resulted in failure, problems, or issues. The issues or problems must be investigated, and the root cause determined. What caused the failure or the problem, and how can it be fixed? Are design changes required to fix the problem? Was the problem caused by incorrect operation or user error? What was not clear or confusing to the users? Did the user interact with the design solution in the way that the design team planned or anticipated? What should be changed to prevent the problem in the future? The changes will become part of the redesign plan.

The problem definition and the design specifications form the set of rubrics and targets for the design testing and assessment. The redesign activity aims to bring the design into alignment and compliance with the design specifications and satisfy the design problem. Sometimes, after assessing the performance of the design solution, it may become necessary to change, adapt, eliminate, or add design specifications. The process is a part of the design process. The designers better understand the problem based on what is learned from testing, checking the design against user expectations and demands, and redesign. Design goals can also change as a result of the redesign process but must be done carefully, documented, and justified.

Testing, assessment, and redesign is an iterative process. For any iterative process, convergence to an optimal solution is desired but not guaranteed. The design team's number of design iterations and progress on design improvements depend directly on the scope and magnitude of the design problem. The capstone design team has limited time and resources to complete the redesign phase, typically one month. With the limited time constraint, the redesign plan should be created carefully and well documented.

If the design has fundamental or inherent flaws, it may not be possible to correct those by redesign only. Sometimes, the designers will have to go back to the beginning, rework the problem definition, and start over the design process.

After each redesign iteration, the design solution should be quickly tested for those aspects that were changed. The redesign iteration continues if the issues still exist or new issues surface. The design team must assess how the changes made affect the solution.

10.1 APPLICATION OF TEST RESULTS

As was discussed in the section on test engineering, the design team prepares a carefully thought out set of tests and lays them out in a test matrix. For capstone design projects, tests are typically conducted to measure the success of the design solution against the selected design specifications. For example, if the weight of a product is specified to be less than 50 lbs, then the working model can be weighed and compared against the specified limit of 50 lbs. If the weight of the working model exceeds the target value, then the design team considers the changes that they can make to reduce weight by changing the design geometry, assembly, materials, etc. If the weight target is met for this iteration round, then the design team moves on to other specifications of interest.

10.2 ADJUSTMENTS TO DESIGN-BUILD

Product or process testing of the working model of the design produces results that should be used in making changes to the design so the build can be achieved or be better. If the build does not satisfy some of the design specifications, then adjustments to the design solution may bring the solution back into compliance with the design specifications.

In some cases, it may not be possible to satisfy all of the design specifications because the system is over-constrained, or it may be that the design specifications were not correctly formulated or too rigid. In such cases, changing the design specifications in play may be the only way that a viable design can be achieved.

The change process to achieve a working design is a normal and expected iterative process that requires the engagement of all stakeholders for the project, particularly when the design specifications may have to be changed.

10.3 MAJOR DESIGN CHANGES

Suppose adjustments to the design solution or changing some of the design specifications do not produce a satisfactory design solution. In that case, it may be necessary to backtrack and make significant changes to the design solution. Making major changes to the design process or solution can be very costly to the design team and set back the project timeline. Setbacks caused by major changes to the design can become irrecoverable.

All stakeholders must be kept informed of the necessity for a major design change. The design team should consider the implications of the significant design change on project milestones and costs. Prior approvals from the sponsors and professor for the course must be obtained with complete explanations, and the team must demonstrate that they understand the implications of the decision.

10.4 OPTIMIZATION

Design projects have a set of objectives (design for X) that are to be considered towards some optimization goal, for example, minimum weight for a product, a minimum time for a process, or minimum cost.

Optimization is determining the design parameter set (specifications) that leads to a maximum objective function such as best performance, for example, maximizing profit and achieving the best performance. Open-ended design problems have many possible design solutions, some of which are identified by the design team. Among the set of solutions identified by the design team, there is at least one solution that produces better performance or a maximum objective function – the input variables to the objective function influence the outcomes. The measurable performance of the design is subject to constraints that bind the problem.

The optimization problem is generally too complicated to achieve a closed-form solution. Numerical methods are typically the preferred method for achieving an optimal design parameter set systematically and efficiently. Typically, the design objective function cannot be expressed in the form of an algebraic function. It may be possible to create a design computer model to achieve a numerical objective function. More practically, testing must be performed on the design to measure the performance function experimentally.

So, we can define the optimization problem as the problem to find a set of design parameters that produce the best measurable performance subject to the constraints specified. The parametric variations may be from continuous or discrete variables as input into the design problem. Continuous variables can be easily discretized for a numerical treatment of the optimization problem. The relationships among design variables may be linear or nonlinear. The optimization problem is typically constrained, but it may contain unconstrained portions that can be treated separately.

The optimization problem is a search problem where a search can be conducted in a region for a local minimum or maximum. Search methods have been studied in-depth and include many traditional mathematical techniques (e.g., bisection method) as well as methods of artificial intelligence such as machine learning, artificial neural networks, and genetic algorithms.

Optimization is an extensive topic, and many excellent textbooks and courses are available and should be considered for further reading by serious engineering designers.

10.5 QUALITY ENGINEERING

The concept of quality is somewhat vague and dependent on the culture and expectations in a particular field or class of products or processes. Many different meanings are associated with the term "quality." Generally, the quality of a product or service is understood as being free of defects, weaknesses, and deficiencies.

Garvin proposed eight critical dimensions or categories: performance, features, reliability, conformance, durability, serviceability, aesthetics, and perceived quality.

Performance describes a product's operating characteristics. For example, a car's performance includes acceleration, braking, handling, top speed, and comfort. Performance is an objective and measurable aspect of quality.

Features are another explicit measure of quality. Features include those aspects explicitly requested by the sponsor and those aspects that the design team decided as an addition. Features distinguish one design solution from another and provide the sponsor with options and features that they may find highly desirable.

Reliability is reflected in the probability that a product or process will not malfunction or fail within a specified period. Reliability can be measured in terms of the mean time between failures or mean time to a first failure. It is a statistical measure, so the measurements must be repeated a sufficient number of times to produce the desired confidence interval. Because reliability measures require operation for a specified period, they are relevant to products or repeating processes.

Conformance measures how the design matches the customary, expected, or established standards. Conformance can be measured as yield (success) rates of the build/manufacturing or user experiences. Conformance is the complement (opposite) to defect rate. Defects are discovered as part of a quality check process during manufacturing or through service and repair records.

Durability is a measure of a product's useful service life. A product or service is less durable if it fails to operate as expected or needs frequent repairs. Failures and repairs add to the net cost of the ownership of the product.

Serviceability is the ease of repairs, speed of completing fixes, and the completeness of restoring the product to a working condition. Sponsors and users demand that the product is restored as quickly as possible after a breakdown or required routine maintenance. Many manufacturers offer a replacement as an option for repairing. A replacement has become a more common option because it increases consumer satisfaction with a product and decreases manufacturer costs.

Aesthetics is an aspect of quality that is subjective. Aesthetics is subject to consumer surveys and preferences, including product look, feel, and sound.

Redesign

Perceived quality is an indirect aspect of a product's value and uses to the customer. A customer's perception of the quality of a product is typically developed through ratings that are now common and found through internet shopping sites. Another way that perceived quality is established is through independent reviews found in technical and trade publications or internet sites, e.g., Consumer Reports.

10.6 COST EVALUATION

Cost is always an essential part of any design project. When the redesign is undertaken and accomplished, the cost is a crucial parameter for improvement. Cost evaluation should be incorporated into the project plan such that resources needed are included and accounted for in the project plan. Cost evaluation must include all resources that have been utilized to accomplish the project, including people, facilities, laboratories, equipment such as 3D printers, and materials acquired. People include the students working on the project, consultants, faculty advisors, mentors, and company (sponsor) staff. A detailed project plan (discussed previously in this book) should capture all of the resources utilized. The report feature in MS Project allows the preparation of reports for cost accounting and evaluation.

The redesign process will impact the costs and financial analysis, including the cost of the redesign activity. The cost evaluation will be part of the decision to redesign.

10.7 RETURN ON INVESTMENT

New design solutions to an existing process or product manufacturing method require the expenditure of funds to implement the new process by acquiring capital equipment. The key question is how long before the savings from the new process will pay back for the initial investment. After that period, the company will be making profits on the investment and the change in the design or process.

The concept of return on investment is based on economic analysis to include the future value of money, similar to a loan. For a loan, the rate of return is the interest rate earned on the outstanding balance for an amortized loan. For example, if a bank lends $50,000 to be repaid in installments over 5 years at a compound interest rate of 5%, the annual payment can be calculated from the following formulas to be $62,500.00 for simple interest and $63,814.08 for compound interest.

$$FV = I \times (1 + (R \times T))$$

For simple interest and from

$$FV = I \times (1 + R)^T$$

For compound interest, where,
 I = initial investment
 R = annual interest rate
 T = number of years

Suppose that the investment of $50,000 in implementing the new design solutions will result in annual savings or additional profits of $5,000 per year. In this case, we need to calculate the number of years it will take to pay off the loan from the savings or profits generated from design changes. We can use the following formula for the payment:

$$P = \frac{R \times PV}{1 - (1 + R)^{-T}}$$

where
P = payment
PV = present value

Plugging in the values for the example:

$$5,000 = \frac{0.05 \times 50,000}{1 - (1 + 0.05)^{-T}}$$

results in $T = 14.21$ years, the duration for return on investment.
The mathematical solution for T can be expressed as

$$T = -\frac{\log\left[1 - \frac{PV \times R}{P}\right]}{\log[1 + R]}$$

10.8 DESIGN OPTIMIZATION

Design optimization is an inherent part of the design process and engineering problem-solving. When creating a design that solves a design problem, many possible solutions are possible. Some solutions are better than others, and there may be a best solution. The purpose of the design optimization process is to find a better solution given a current design solution.

Optimization methods are well developed, and many mathematical and numerical methods exist for optimization, mainly when a mathematical model for the performance function of the design solution can be created. Linear optimization methods can be used for the simpler parts of the design. More involved nonlinear mathematical models for design optimization are applicable in some special cases but are generally difficult for undergraduate students to use.

Other practical design optimization methods are more suitable for capstone design projects, including testing and improvements, trial-and-error, numerical algorithms, and intuitive methods.

Testing and user trials were described previously. Design improvements can be based on weaknesses, deficiencies, failures, or errors discovered during testing and user trials. Such changes are a form of design optimization towards a better design solution. A design change is a direct result of identifying specific issues. This form of design optimization is experiential and evolutionary.

Redesign

Designs can also be optimized by trial and error. The idea is to make a design change and check performance against design specifications. If the changed design performs better than before, then keep the change. If it performs more poorly, then reject the change and move on to the next idea. The change ideas can come from many different sources, including intuition, educated guesses, sponsor suggestions, and mentor suggestions. Engineers in the profession use trial and error frequently based on their intuition and suggestions by other experts.

Another way to optimize the design is using numerical algorithms and methods. An objective function is defined for each design feature to represent the design. The objective function can be represented as:

$$\vec{f}(\vec{x}) = \begin{bmatrix} f_1(x_1, x_2, \ldots, x_n) \\ f_2(x_1, x_2, \ldots, x_n) \\ \vdots \\ f_m(x_1, x_2, \ldots, x_n) \end{bmatrix}$$

The objective function is a vector function with each component representing an aspect of the design such as cost, performance, weight, strength, thermal output, a factor of safety. The parameters $(x_1, x_2, \ldots, x_n)$ represent the design parameters under designer control that can be changed such as choice of material, dimensions, configuration, loads, temperatures, flow rates, acceleration, speeds, and process activity time.

The number of design parameters to be manipulated should be kept small to make the optimization problem manageable. Also, the characteristics of the objective function may be nonlinear with respect to the design parameters to be varied. The optimization process can be better managed if the parametric variations are also kept small to treat the optimization problem as quasilinear.

11 Closing Out the Project and Documentation

The completion of the capstone design project is a significant and momentous accomplishment towards the completion of a bachelor of science degree in engineering. Sponsors look forward to receiving the full description of the design solutions achieved by student teams in the form of a final presentation of results and a detailed final design report. Students are proud of their accomplishments and look forward to sharing their engineering design work with stakeholders, including the sponsors. Faculty and mentors who have worked with the student design team and provided guidance and advice to the students look forward to seeing the final product and the design solution.

The culmination of the capstone design projects should be a celebration of the many accomplishments of the team and the final design solution reporting to the community. The reporting out of the accomplishments of the design work can take many different forms such as a symposium, conference, poster session, and design showcase. The design team will also be preparing their final design report to transmit to the sponsor and the professor as part of the capstone course requirements. Other documents such as a user's guide, programming manual, or a maintenance manual may also be prepared for transmission to the sponsor and as part of the design documentation.

Planning and participation by all students, faculty, mentors, advisors, and sponsors will contribute to a successful event and culmination of the capstone design projects.

11.1 PLANNING A FINAL CAPSTONE EVENT

Planning for the event is an essential activity for the successful completion of the capstone design experience. Essential steps in the planning of the event are outlined in the following.

11.1.1 Estimating the Audience Size and Selecting a Venue

The audience for the final capstone event should include all of the students in the capstone program, faculty, mentors, sponsors, teaching assistants, and advisors. In addition to the stakeholders mentioned, it is a good idea also to invite the junior class (who will be in capstone the following year), alumni who are engaged with the engineering program, prospective sponsors, members of the industrial advisory board for the engineering program and the college, and family members and friends of the students in the capstone. The audience size can be estimated by sending out invitations a couple of months before the event and asking for responses to more accurately estimate the audience size. The audience size will determine the size of the venue that will need to be reserved and prepared. Most campuses have facilities or classrooms that are sufficiently large to accommodate up to several hundred people attending the event. The facility or the classroom needs to be reserved for the duration of the event.

11.1.2 Select a Date

A date must be selected for the event that allows all students and as many of the other stakeholders to be able to participate. A good option is to schedule the event at the very end of the academic semester, perhaps on the last class day. The duration of the event will impact the date directly. Suppose the event is a symposium, where each design team presents their work in a 20–30-minute presentation. In that case, the duration is determined by the number of the capstone projects times the duration of each presentation plus time for registration and breaks throughout the event. If the event is planned in a symposium format, it will take several hours to a full day to get through all the presentations.

A design showcase can be scheduled for a fixed duration such as 2 or 3 hours. The design teams can present their work in a poster format, including any physical prototype they built during a design showcase. The advantage of the design showcase format is that it can accommodate a medium to a large number of students and projects. A 3-hour duration for the design showcase can accommodate 100–150 students and 30–40 projects.

11.1.3 Select a Format

The logistics of class size and the number of projects will likely drive the decision as to the event's format. For small class sizes (up to 40 students or 10 teams), a conference or symposium format may be possible in half a day time frame.

For large classes (>40) and many projects (>10), a design showcase with a poster presentation format is more time-efficient for the students and the attendees.

Figure 11.1 shows the format of a typical design showcase. The event shown was held at the University of Rhode Island in April 2018.

11.1.4 Create a Program

Once a format for the event has been decided, a program should be created to describe the event and the projects that will be presented. For a seminar format, list the time and location for each project presentation, the project title, the names of students, and the names of the sponsor and the faculty advisor for that project. The symposium schedule will include information for attendees and times and locations for all of the presentations.

For a showcase format, create a list of all projects that will be on display at the event and names of students who will be presenting and their sponsor and faculty advisor. The program should be available to attendees at the event in paper form and also on a website dedicated to the event.

Closing Out the Project and Documentation

FIGURE 11.1 Capstone design showcase format.

11.1.5 Promote the Event and Invite People

The event should be promoted by creating materials and messages that can be easily distributed by e-mail and social media. Sponsors should be invited at least a month before the event. Faculty, mentors, and advisors should also be invited early to allow them time to schedule the event in their calendars. The junior class should also be invited so they can experience the capstone process and learn from the design projects for when they become seniors in capstone the following year. The event should also be announced and invitations sent to alumni groups and advisory boards for the engineering program and the college of engineering. The promotion of the final engineering capstone presentations is essential to the vitality and the continuity of the sponsored industry projects for future years.

Social media platforms such as LinkedIn, Facebook, and Instagram can be used to promote the event and invite interested individuals to attend the event and learn about the accomplishments of the student teams.

Distribute the event schedule of presentations and the program to the attendees who have responded to attend the event.

11.1.6 Invite Media

It is crucial to promote the event by inviting the media such as the school news, local television stations, and radio stations to the event. Also, photographs and videos should be taken during the event to document the event and document the events. Since the students will be attending and are dressed professionally for the event, this presents an opportunity to take photos of the teams along with their professional poster presentations and prototypes of their products. Because the capstone design symposium or showcase is scheduled before the teams complete their final report, the team photos taken during the event can be incorporated on the cover of the design reports. See examples of this in the sample design reports provided in the online files for this book.

11.2 DESIGN PRESENTATION TO SPONSOR

An excellent closure to capstone design projects with industrial sponsors is to schedule a final meeting with the sponsor contact and their management and include the capstone program director or professor to make a 30–60 minute in-depth presentation about the project and what was accomplished. Industry partners appreciate such end-of-year meetings and project finalization and handover of final reports, supporting files, and physical prototypes (if appropriate).

Sufficient lead time must be given to schedule such a meeting. Typically, 30 days before the meeting is a good guideline for requesting the meeting and scheduling it. Typically, the final company meeting will be scheduled after all of the capstone course requirements deadlines and milestones have been met, i.e., after the end of the academic semester. The meeting should be scheduled at the company location to allow their management and engineering teams to participate. Students working on the project also benefit from participating in this final presentation. It is similar to the company's typical project meetings, and the experience is invaluable for all participating.

Bibliography

Adams, D.F., Odom, E.M., 1987. Testing of single fibre bundles of carbon/carbon composite materials. *Composites* 18, 381–385. https://doi.org/10.1016/0010-4361(87)90362-4.

Al-Thani, S.B.J., Abdelmoneim, A., Cherif, A., Moukarzel, D., Daoud, K., 2016. Assessing general education learning outcomes at Qatar University. *Journal of Applied Research in HE* 8, 159–176. https://doi.org/10.1108/JARHE-03-2015-0016.

Applied Imagination - Wikipedia [WWW Document], n.d. https://en.wikipedia.org/wiki/Applied_Imagination (accessed 7.12.20).

Atadero, R.A., Rambo-Hernandez, K.E., Balgopal, M.M., 2015. Using social cognitive career theory to assess student outcomes of group design projects in statics: SCCT and student outcomes of statics projects. *Journal Engineering Education* 104, 55–73. https://doi.org/10.1002/jee.20063.

Benavides, E., 2011. *Advanced Engineering Design*. Woodhead Publishing. https://proquest.safaribooksonline.com/book/mechanical-engineering/9780857090935.

Buchsbaum, A., Rey, C., n.d. Jean-Yves Beziau Federal University of Rio de Janeiro Rio de Janeiro, RJ Brazil 623.

Buzzetto-More, N.A., Julius Alade, A., 2006. Best practices in e-assessment. *JITE: Research* 5, 251–269. https://doi.org/10.28945/246.

Caple, M., Wild, J., Maslen, E., Nagel, J., 2015. Design of a controller for a hybrid bearing system, in: 2015 Systems and Information Engineering Design Symposium. *Presented at the 2015 Systems and Information Engineering Design Symposium*, IEEE, Charlottesville, VA, USA, pp. 236–239. https://doi.org/10.1109/SIEDS.2015.7116980.

Carnevalli, J.A., Miguel, P.A.C., Calarge, F.A., 2010. Axiomatic design application for minimising the difficulties of QFD usage. *International Journal of Production Economics* 125, 1–12. https://doi.org/10.1016/j.ijpe.2010.01.002.

Carter, R., Strader, T., Rozycki, J., Root, T., , 2015. Cost structures of information technology products and digital products and services firms: Implications for financial analysis. *JMWAIS* 1, 5–19. https://doi.org/10.17705/3jmwa.00002.

CATIA [WWW Document], n.d. https://www.3ds.com/products-services/catia/?wockw=card_content_cta_1_url%3A%22https%3A%2F%2Fblogs.3ds.com%2Fcatia%2F%22.

Cordon, D., Clarke, E., Westra, L., Allen, N., Cunnington, M., Drew, B., Gerbus, D., Klein, M., Walker, M., Odom, E.M., Rink, K.K., Beyerlein, S.W., 2002. Shop orientation to enhance design for manufacturing in capstone projects, in: 32nd Annual Frontiers in Education. *Presented at the Conference on Frontiers in Education*, IEEE, Boston, MA, USA, pp. F4D-6-F4D-11. https://doi.org/10.1109/FIE.2002.1158229.

Council on Higher Education (South Africa), 2011. *Work-Integrated Learning: Good Practice Guide*. Council on Higher Education, Pretoria, South Africa.

Creativity Unbound [WWW Document], n.d. FourSight. https://foursightonline.com/product/creativity-unbound/ (accessed 8.16.20).

Criteria for Accrediting Engineering Programs, 2019–2020 | ABET [WWW Document], n.d. https://www.abet.org/accreditation/accreditation-criteria/criteria-for-accrediting-engineering-programs-2019-2020/ (accessed 6.10.20).

Davis, D.C., Gentili, K.L., Trevisan, M.S., Calkins, D.E., 2002. Engineering design assessment processes and scoring scales for program improvement and accountability. *Journal of Engineering Education* 91, 211–221. https://doi.org/10.1002/j.2168-9830.2002.tb00694.x.

Deveci, T., Nunn, R., 2018. COMM151: A PROJECT-BASED COURSE TO ENHANCE ENGINEERING STUDENTS' COMMUNICATION SKILLS. *Journal of Teaching English for Specific and Academic Purposes* 6, 027. https://doi.org/10.22190/JTESAP1801027D.

Diefes-Dux, H.A., Moore, T., Zawojewski, J., Imbrie, P.K., Follman, D., 2004. A framework for posing open-ended engineering problems: Model-eliciting activities, in: 34th Annual Frontiers in Education, 2004. FIE 2004. *Presented at the 34th Annual Frontiers in Education*, 2004. FIE 2004, IEEE, Savannah, GA, USA, pp. 455–460. https://doi.org/10.1109/FIE.2004.1408556.

Dollar, A.M., Kerdok, A.E., Diamond, S.G., Novotny, P.M., Howe, R.D., 2005. Starting on the right track: Introducing students to mechanical engineering with a project-based machine design course, in: Innovations in Engineering Education: Mechanical Engineering Education, Mechanical Engineering/Mechanical Engineering Technology Department Heads. *Presented at the ASME 2005 International Mechanical Engineering Congress and Exposition*, ASMEDC, Orlando, Florida, USA, pp. 363–371. https://doi.org/10.1115/IMECE2005-81929.

Duarte, B.B., de Castro Leal, A.L., de Almeida Falbo, R., Guizzardi, G., Guizzardi, R.S.S., Souza, V.E.S., 2018. Ontological foundations for software requirements with a focus on requirements at runtime. *Applied Ontology* 13, 73–105. https://doi.org/10.3233/AO-180197.

Dynn, C.L., Agogino, A.M., Eris, O., Frey, D.D., Leifer, L.J., 2006. Engineering design thinking, teaching, and learning. *IEEE Engineering Management Review* 34, 65–65. https://doi.org/10.1109/EMR.2006.1679078.

Eberle, B., 2008. *Scamper: Creative Games and Activities for Imagination Development.* Prufrock Press, Austin, TX.

Egelhoff, C., Odom, E., 2001. Advanced learning made as easy as ABC: An example using design for fatigue of machine elements subjected to simple and combined loads, in: 31st Annual Frontiers in Education Conference. Impact on Engineering and Science Education. Conference Proceedings (Cat. No.01CH37193). *Presented at the 31st Annual Frontiers in Education Conference. Impact on Engineering and Science Education*, IEEE, Reno, NV, USA, pp. T4B-7-T4B-12. https://doi.org/10.1109/FIE.2001.963934.

Egelhoff, C.F., Odom, E.M., 1999. Machine design: Where the action should be, in: FIE'99 Frontiers in Education. 29th Annual Frontiers in Education Conference. Designing the Future of Science and Engineering Education. Conference Proceedings (IEEE Cat. No.99CH37011. *Presented at the IEEE Computer Society Conference on Frontiers in Education*, Stripes Publishing L.L.C, San Juan, Puerto Rico, p. 12C5/13-12C5/18. https://doi.org/10.1109/FIE.1999.841644.

Egelhoff, C.J., Odom, E.M., Wiest, B.J., 2010. Application of modern engineering tools in the analysis of the stepped shaft: Teaching a structured problem-solving approach using energy techniques, in: 2010 IEEE Frontiers in Education Conference (FIE). *Presented at the 2010 IEEE Frontiers in Education Conference (FIE)*, IEEE, Arlington, VA, USA, pp. T1C-1-T1C-6. https://doi.org/10.1109/FIE.2010.5673504.

Farid, A.M., Suh, N.P. (Eds.), 2016. *Axiomatic Design in Large Systems*. Springer International Publishing, Cham. https://doi.org/10.1007/978-3-319-32388-6.

Griffin, P.M., Griffin, S.O., Llewellyn, D.C., 2004. The impact of group size and project duration on capstone design. *Journal of Engineering Education* 93, 185–193. https://doi.org/10.1002/j.2168-9830.2004.tb00805.x.

Howe, S., Rosenbauer, L., Poulos, S., n.d. 2015 capstone design survey – Initial results, in: *Proceedings of the 2016 Capstone Design Conference*, Columbus, OH, vol. 4, pp. 6–8.

Leaman, E.J., Cochran, J.R., Nagel, J.K., 2014. Design of a two-phase solar and fluid-based renewable energy system for residential use, in: 2014 Systems and Information

Bibliography

Engineering Design Symposium (SIEDS). *Presented at the 2014 Systems and Information Engineering Design Symposium (SIEDS)*, IEEE, Charlottesville, VA, USA, pp. 193–197. https://doi.org/10.1109/SIEDS.2014.6829917.

Morell, L., 2015. Disrupting engineering education to better address societal needs, in: 2015 International Conference on Interactive Collaborative Learning (ICL). *Presented at the 2015 International Conference on Interactive Collaborative Learning (ICL)*, IEEE, Firenze, Italy, pp. 1093–1097. https://doi.org/10.1109/ICL.2015.7318184.

Nagel, J., 2013. Guard cell and tropomyosin inspired chemical sensor. *Micromachines* 4, 378–401. https://doi.org/10.3390/mi4040378.

Nagel, J.K.S., Liou, W.F., 2012. Hybrid manufacturing system design and development, in: Abdul Aziz, F. (Ed.), *Manufacturing System*. InTech. https://doi.org/10.5772/35597.

Nagel, J.K.S., Nagel, R.L., Eggermont, M., 2013. Teaching biomimicry with an engineering-to-biology thesaurus, in: Volume 1: 15th International Conference on Advanced Vehicle Technologies; 10th International Conference on Design Education; 7th International Conference on Micro- and Nanosystems. *Presented at the ASME 2013 International Design Engineering Technical Conferences and Computers and Information in Engineering Conference*, American Society of Mechanical Engineers, Portland, Oregon, USA, p. V001T04A017. https://doi.org/10.1115/DETC2013-12068.

Nagel, J.K.S., Nagel, R.L., Stone, R.B., 2011. Abstracting biology for engineering design. *IJDE* 4, 23. https://doi.org/10.1504/IJDE.2011.041407.

Nagel, J.K.S., Nagel, R.L., Stone, R.B., McAdams, D.A., 2010. Function-based, biologically inspired concept generation. *AIEDAM* 24, 521–535. https://doi.org/10.1017/S0890060410000375.

Nagel, J.K.S., Stone, R.B., 2012. A computational approach to biologically inspired design. *AIEDAM* 26, 161–176. https://doi.org/10.1017/S0890060412000054.

Nagel, J.K.S., Stone, R.B., McAdams, D.A., 2010. An engineering-to-biology thesaurus for engineering design, in: Volume 5: 22nd International Conference on Design Theory and Methodology; Special Conference on Mechanical Vibration and Noise. *Presented at the ASME 2010 International Design Engineering Technical Conferences and Computers and Information in Engineering Conference*, ASMEDC, Montreal, Quebec, Canada, pp. 117–128. https://doi.org/10.1115/DETC2010-28233.

Odom, E.M., Adams, D.F., 1990. Failure modes of unidirectional carbon/epoxy composite compression specimens. *Composites* 21, 289–296. https://doi.org/10.1016/0010-4361(90)90343-U.

Odom, E.M., Beyerlein, S.W., Tew, B.W., Smelser, R.E., Blackketter, D.M., 1999. Idaho Engineering Works: A model for leadership development in design education, in: FIE'99 Frontiers in Education. 29th Annual Frontiers in Education Conference. Designing the Future of Science and Engineering Education. Conference Proceedings (IEEE Cat. No.99CH37011. *Presented at the IEEE Computer Society Conference on Frontiers in Education*, Stripes Publishing L.L.C, San Juan, Puerto Rico, p. 11B2/21-11B2/24. https://doi.org/10.1109/FIE.1999.839217.

Odom, E.M., Egelhoff, C.J., 2011. Teaching deflection of stepped shafts: Castigliano's theorem, dummy loads, heaviside step functions and numerical integration, in: 2011 Frontiers in Education Conference (FIE). *Presented at the 2011 Frontiers in Education Conference (FIE)*, IEEE, Rapid City, SD, USA, pp. F3H-1-F3H-6. https://doi.org/10.1109/FIE.2011.6143039.

Online, S.B., n.d. Advanced Engineering Design [WWW Document]. https://learning.oreilly.com/library/view/advanced-engineering-design/9780857090935/ (accessed 6.6.20).

ORACLE PRIMAVERA P6 SOFTWARE [WWW Document], n.d. https://globalpm.com/products/oracle-primavera-p6-software/.

Osborne, A.F., 1953. *Applied Imagination; Principles and Procedures of Creative Thinking.* (Book, 1953) [WorldCat.org] [WWW Document]. https://www.worldcat.org/title/applied-imagination-principles-and-procedures-of-creative-thinking/oclc/641122686 (accessed 7.12.20).

Pahl, A.K., Newnes, L., McMahon, C., 2007. A generic model for creativity and innovation: Overview for early phases of engineering design. *JDR* 6, 5. https://doi.org/10.1504/JDR.2007.015561.

Pembridge, J.J., Paretti, M.C., 2019. Characterizing capstone design teaching: A functional taxonomy. *Journal of Engineering Education* 108, 197–219. https://doi.org/10.1002/jee.20259.

Ribeiro, A.L., Bittencourt, R.A., 2018. A PBL-based, integrated learning experience of object-oriented programming, data structures and software design, in: 2018 IEEE Frontiers in Education Conference (FIE). *Presented at the 2018 IEEE Frontiers in Education Conference (FIE),* IEEE, San Jose, CA, USA, pp. 1–9. https://doi.org/10.1109/FIE.2018.8659261.

Ritter, S.M., Mostert, N., 2017. Enhancement of creative thinking skills using a cognitive-based creativity training. *Journal of Cognitive Enhancement* 1, 243–253. https://doi.org/10.1007/s41465-016-0002-3.

Robert Lee Hotz., Print, 1999. Mars Probe Lost Due to Simple Math Error [WWW Document]. Los Angeles Times. https://www.latimes.com/archives/la-xpm-1999-oct-01-mn-17288-story.html (accessed 8.16.20).

Romaniuk, R.S., 2012. Communications, multimedia, ontology, photonics and internet engineering 2012. *International Journal of Electronics and Telecommunications* 58, 463–478. https://doi.org/10.2478/v10177-012-0061-z.

Salustri, F.A., Eng, N.L., Weerasinghe, J.S., 2008. Visualizing information in the early stages of engineering design. *Computer-Aided Design and Applications* 5, 697–714. https://doi.org/10.3722/cadaps.2008.697-714.

Stroble, J.K., Stone, R.B., Watkins, S.E., 2009. Assessing how digital design tools affect learning of engineering design concepts, in: Volume 8: 14th Design for Manufacturing and the Life Cycle Conference; 6th Symposium on International Design and Design Education; 21st International Conference on Design Theory and Methodology, Parts A and B. *Presented at the ASME 2009 International Design Engineering Technical Conferences and Computers and Information in Engineering Conference,* ASMEDC, San Diego, California, USA, pp. 467–476. https://doi.org/10.1115/DETC2009-86708.

Stroble, J.K., Stone, R.B., Watkins, S.E., 2009. An overview of biomimetic sensor technology. *Sensor Review* 29, 112–119. https://doi.org/10.1108/02602280910936219.

Swartwout, M., Kitts, C., Twiggs, R., Kenny, T., Ray Smith, B., Lu, R., Stattenfield, K., Pranajaya, F., 2008. Mission results for Sapphire, a student-built satellite. *Acta Astronautica* 62, 521–538. https://doi.org/10.1016/j.actaastro.2008.01.009.

The Gantt chart, a working tool of management: Clark, Wallace, 1880–1948: Free Download, Borrow, and Streaming [WWW Document], n.d. Internet Archive. https://archive.org/details/cu31924004570853 (accessed 7.6.20).

Todd, R.H., Magleby, S.P., Sorensen, C.D., Swan, B.R., Anthony, D.K., 1995. A survey of capstone engineering courses in North America. *Journal of Engineering Education* 84, 165–174. https://doi.org/10.1002/j.2168-9830.1995.tb00163.x.

Truong, H.T.X., Odom, E.M., Egelhoff, C.J., Burns, K.L., 2011. Using modern engineering tools to efficiently solve challenging engineering design problems: Analysis of the stepped shaft, in: *ASEE Middle Atlantic Regional Conference,* April 29-30, 2011, Farmingdale State College, SUNY.

Tuckman, B.W., Jensen, M.A.C., 1977. Stages of small-group development revisited. *Group & Organization Studies* 2, 419–427. https://doi.org/10.1177/105960117700200404.

Wenhao Huang, D., Diefes-Dux, H., Imbrie, P.K., Daku, B., Kallimani, J.G., 2004. Learning motivation evaluation for a computer-based instructional tutorial using ARCS model of motivational design, in: 34th Annual Frontiers in Education, 2004. FIE 2004. *Presented at the 34th Annual Frontiers in Education, 2004.* FIE 2004, IEEE, Savannah, GA, USA, pp. 65–71. https://doi.org/10.1109/FIE.2004.1408466.

Why ABET Accreditation Matters | ABET [WWW Document], n.d. https://www.abet.org/accreditation/what-is-accreditation/why-abet-accreditation-matters/ (accessed 6.10.20).

Williams, C.B., Gero, J., Lee, Y., Paretti, M., 2010. Exploring spatial reasoning ability and design cognition in undergraduate engineering students, in: Volume 6: 15th Design for Manufacturing and the Lifecycle Conference; 7th Symposium on International Design and Design Education. *Presented at the ASME 2010 International Design Engineering Technical Conferences and Computers and Information in Engineering Conference*, ASMEDC, Montreal, Quebec, Canada, pp. 669–676. https://doi.org/10.1115/DETC2010-28925.

Xiao, A., Park, S.S., Freiheit, T., 2011. A comparison of concept selection in concept scoring and axiomatic design methods. *Proceedings of the Canadian Engineering Education Association (PCEEA).* https://doi.org/10.24908/pceea.v0i0.3769.

Yusof, K. Mohd., Hassan, S. A. H. S., Jamaludin, M. Z., Harun, N. F., 2012. Cooperative Problem-Based Learning (CPBL): Framework for integrating cooperative learning and problem-based learning. *Procedia - Social and Behavioral Sciences* 56, 223–232.

zotero / Insert Items from Amazon [WWW Document], n.d. http://zotero.pbworks.com/w/page/5511972/Insert%20Items%20from%20Amazon (accessed 8.16.20).

Index

3D printers 43, 64, 95, 153, 167, 168, 179
3D printing 42, 118, 152, 160, 162, 167, 168

AAAS 13
AAES 14
Abaqus 118
ABET 6, 8–10, 16
ABS 101, 153, 155, 160, 161
accessibility/accessible 36, 124
account 36, 37, 42, 70, 172, 174
accountability 84
accounting 56, 179
Accreditation Board for Engineering and Technology 6
accreditation 6, 8–10, 14, 16
ACERS 14
Acetal 155, 161
ACM 14
Acoustical Society of America 14
acoustics 14
acrylic 161
ACT 11, 49, 111
acupuncture 13
adaptability 92
additive 167
adhesives 153, 155
adjourning 29
advertising regulation 12
advertising 12, 133
advisory boards 185
advisory 183, 185
aeronautics 12, 14, 85
aerospace 15
aesthetics 3, 15, 17, 18, 20, 22, 109, 112, 178
affordability 106
age 68
agencies 11, 12
agenda 30–32
aggregated 29
agricultural 14
AIAA 14
AIChE 14
AIME 14
airports 11
alertability 125
Alex Osborne 64, 86
algorithms/algorithmic 77, 90, 95, 178, 180, 181
alloys 155, 159, 160
Altshuller, Genrich 86, 90
Alumina 155

aluminum 155, 158, 159, 164
alumni 183, 185
American Academy of Environmental Engineers and Scientists 14
American Association for the Advancement of Science 13
The American Ceramics Society 14
American Institute of Aeronautics and Astronautics 14
American Institute of Chemical Engineers 14
American Nuclear Society 14
American Society for Engineering Education 14
American Society of Agricultural and Biological Engineers 14
American Society of Civil Engineers 14
American Society of Mechanical Engineers 14
amortized 179
Android 36, 152
angle 158, 159
annotate/annotations 129, 134, 135, 139, 140
annual 179, 180
anodized 159
ANS 14
ANSYS 94
appendix 139, 145
Appft 74
Arduino 101, 164, 167
ARIZ 90, 91
arrows 44, 51, 135
art 3, 139, 143
artificial intelligence 90, 178
ASA 14
ASABE 14
ASCE 14
ASEE 14, 131
ASHRAE 14
ASME 14, 117, 139, 141, 145, 147
assemblies 173
The Association for Computing Machinery 14
ASTM 14, 117, 144
astronautics 12, 14
asymmetry 93
athletic 27
atmosphere 94
atomic 156
attitude 23, 27–29, 48, 84, 129
attributes 16, 24, 41, 48, 64, 76, 77, 93
AutoCAD 43
automation 90, 92, 165
autonomy/autonomous 100, 101

193

auto-scheduled 44
availability 19, 42, 49, 57, 65, 163
axioms/axiomatic 18, 90

backtrack 177
backward 41
balancing 56
balsa 154
baseline 117, 130, 140, 145
basis 107, 110, 113
beam 158, 159
behavior 64, 65, 84, 117
benchmarking 95
bench-scale 144
bias 65, 73, 83
bibliography 71, 88
Bill of Materials 117, 154, 173
Biomedical Engineering Society 14
Blessing in disguise 93
block 84, 165
blogs 71
blueprints 117
BMES 14
bolts 153, 163
BOM 117, 154, 173
boolean 75
boosted 94
bottlenecks 29, 39, 56
brainstorm/brainstormed/
  brainstorming 17, 63, 64, 70,
  71, 73, 76, 84–88, 110, 116, 133
brasses 155, 160
breakthrough 64
brick 155
brightness 126
brilliant 83, 109
brittle 157, 158
brochure 114, 131, 118
browser 36
brushless 164
bubble 156
budget 6, 39, 100, 110, 152–154, 175
build 4, 7, 19, 20, 29, 64, 65, 85, 88,
  95, 100, 109–111, 143, 149,
  151–154, 163, 167, 172, 173,
  176, 178
build-plan 151
build-space 152
build-test-redesign 19
bureaucratic 173

cable 12, 165, 168
CAD 35, 43, 117, 118, 167, 168, 173
calendar 16, 30, 36, 41, 49, 52, 56,
  59, 117
calibration 64
callouts 135, 145

capstone 3, 6, 9, 11, 16–27, 29–33, 35, 39–45, 48,
  50, 51, 54–57, 63, 64, 68, 69, 72, 77,
  78, 80, 82–85, 88, 90, 107, 109–113,
  116, 117, 119, 128, 129, 131, 136, 137,
  139, 142, 151, 153–155, 160, 163, 164,
  172–174, 176, 180, 183–186
carbide 155
cartesian 167
cash flow 59, 61
cast iron 155, 157
CATIA 43
cause and effect 88, 165
CDR 18, 107, 108
ceramics 14, 155–159
certification 12, 13, 27, 95
CFD 118
channel 158, 159
characteristics 16, 100, 101, 103, 105, 141, 147,
  155, 168, 178, 181
cheap/cheaper 93, 159, 163
check-points 41
Chrome 36, 152
Cisco 36
Civil Rights Act 11
claims 76
clamps 153
classification 73, 74, 76–78
clients 27, 45, 56
clipart 135
closed-form 177
closing 183
closure 186
cloud 36, 119, 152
cloud-based 43
clues 72
coatings 155
codes 11–13, 15–17, 19, 20, 35, 82, 94, 95, 118,
  139, 143
coding 42, 133
cognitive 84
cognitive-based 85
cold-finished 159
collaborative 9, 29, 48
colleges 8, 23
combine 86, 87, 163
comfort 85, 178
commercialization 144
communications 4, 7, 9, 11, 13, 16, 23, 26, 27,
  29, 30, 33–36, 35, 39, 41, 49, 51, 84,
  85, 107, 113, 114, 115, 116, 119, 131,
  136, 168
community-based 9
competitions 12, 16, 40, 83, 88, 102, 106, 117,
  118, 138, 139, 143
competitive 4, 7, 80, 99, 107, 132, 138,
  143, 173
competitors 3, 17, 18, 25, 69, 70, 102, 106

Index

complex 9, 11, 12, 23, 30, 39, 42, 88, 90, 94, 95, 113, 137, 154, 165
complexity 19, 24, 92, 163
compliance 12, 16, 19, 82, 144, 176
composite 94, 155
compound interest 179
compression 157
computational 95, 117, 167
compute-intensive 167
Computing Sciences Accreditation Board 14
COMSOL 94, 118
conceive/conceiving 1, 17, 18, 83
concepts 3, 4, 6, 7, 17, 18, 22, 40, 64, 81, 83, 88, 94–96, 100, 107, 110, 114, 116–118, 137, 138, 143, 152
concrete 155
conductivity 160
conductors 158
conferences/conferencing 35, 36, 49, 51, 69, 114, 118, 131, 136, 141, 146, 147, 183, 184
confidence 39, 178
confidentiality 28, 33, 34
conflicts 28, 42, 48, 134, 152
conform 16
conformance 178
consensus 29, 48, 96, 102
consortium 16
constraints 3, 4, 7, 9, 18, 39, 48, 63, 78, 90, 107, 143, 144, 175, 177
consultants 118, 139, 144, 179
consultation 80, 118
Consumer Product Safety Commission 11
consumers 11, 12, 13, 16, 18, 178, 179
contemporary 78
continuous improvement 8, 10
contractual 6
controller 167
convergence 76, 176
coops 21, 24, 26, 27, 172
copper 155, 160
copying 93
corporations 72
correlation 102, 106, 156
corrosion 158, 159
counseling 23
counteraction 93
counterweight 93
coursework 26, 28
CPC 73, 74, 76, 78
CPSC 11, 12
creatively/creativity 29, 30, 83, 84, 91, 129
CRITERIA 2000 6
criterion/criteria 6, 8, 9, 16, 78, 96, 97, 99, 102, 104, 111, 116, 138, 143
critical thinking 22, 41, 63, 113, 118

critical 17, 18, 22, 33, 40, 41, 52–55, 59, 63, 80, 84, 85, 95, 107, 109, 113, 114, 116–118, 132, 135, 136, 144, 153, 163, 167, 175, 178
criticisms 64, 86
critiques 18, 107, 108, 129
cross-referenced 117
Crump, Lisa 167
Crump, Scott 167
CSAB 14
CTIS 136, 137
cultural 9, 30
curiosity 29, 30
curricula/curriculum 6, 8–10, 16, 21, 22, 26, 41
curves 93
cushion 93
customer 6, 7, 20, 32, 65, 66, 69, 78–81, 90, 97–100, 102, 103, 106, 110, 111, 138, 143, 175, 179
cyber-security 13

dangerous 135, 165
deadlines 30, 32, 57, 117, 151, 175, 186
decommissioning 144
decomposition 39, 42, 88–90
defects 157, 178
definitions 3, 76, 80
deformation 157
delays 173
delete 87
Delrin 161
demand 8, 100, 101, 113, 138, 139, 143, 144, 176, 178
demanded qualities 100
demographics 68
density 155, 156, 158
dependencies 40, 42, 50, 51, 53
design binder 113, 114, 116, 117
design-build 153, 172, 176
design-domain-dependent 94
design showcase 20, 114, 183–185
DFM 163
DFMA 163
diamond 155
disgruntled 28
divide-and-conquer 42
docs 36, 87
dollar equivalent 139, 144
double-checked 85
dress 27, 28
Dropbox 36, 119
ductile 157, 158
dynamism 93
dysfunction 29
delivery 6, 11, 34, 119
decision-making 6, 18, 29, 95–97, 113, 118
deficiencies 9, 32, 178, 180

documentation/documenting 17, 20, 41, 112–114, 118, 129, 130, 155, 173, 183
databases 18, 25, 42, 63, 70–75
down-select 18, 40
diverse/diversity 23, 24, 26, 29, 30, 49, 68

Eberle, Bob 86
ECC 131
Engineering Accreditation Committee (EAC) 8
engineering characteristic 97, 102
Engineer's Council for Professional Development (EPCD) 6
Environmental Protection Agency (EPA) 11, 12
epoxy 155
Equal Employment Opportunity Commission (EEOC) 11
essential qualities 27
ETCFE 161
evolution 91, 113, 114, 129, 144, 151
Excel 36, 95
excessive 93
extruder 167, 170
economic 9, 13, 118, 144, 179
ergonomics 17, 18, 20, 77, 143, 144
ethics 17, 20, 49

fabrics/fabrication 154, 165
Facebook 185
FaceTime 33
face-to-face 33, 34
failure 20, 24, 28, 45, 84, 85, 110, 144, 157, 165, 175, 178
far-fetched 64
fasteners 153, 163, 164
FCC 11, 144
FDM 167
FDR 114, 142
FED 12
Federal Aviation Administration (FAA) 11, 12
Federal Deposit Insurance Corporation (FDIC) 12
Federal Reserve System 12
FFDM 167
fiberglass 155
filament 167
file-sharing 36, 137
financial 13, 18, 20, 22, 23, 30, 42, 45, 49, 50, 56, 109, 112, 118, 139, 144, 179
finite element 94, 139, 143
flowchart 4, 6, 76, 143, 139, 165
foams 156–159
Food and Drug Administration (FDA) 12
forecasting 138, 143
FTC 12
funding 6, 16, 19, 41, 57, 144
fused 167

galvanized 158
GANTT chart 33, 50–55, 111, 117, 138, 144, 145
g-code 168
geometry 87, 95, 167, 168, 176
Glasgow 96
glass 155, 167
glues 155, 160
Gmail 36, 37
Google 25, 36, 37, 66, 68, 70, 72, 73, 87, 88, 108, 119
GPA 156, 157
G-Suite 37, 66
gussets 158

hangouts 36, 37
HDPE 155, 161
hexagon 159
humidity 82
hydraulics 93

IEEE-CS 14
IISE 15
imagination 64, 83, 84
ImechE 15
INCOSE 14
industrial advisory board 183
industrial 15, 183, 186
infill 172
information sheet 131, 136, 137
innovation/innovative 13, 14, 64, 84, 129
inspirational 49
Instagram 185
installability 77
installments 179
Institute of Electrical and Electronics Engineers (IEEE) 14, 16
Institute of Transportation Engineers (ITE) 15
Institution of Engineering and Technology (IET) 14
intellectual property 72, 73, 114
intermediary 93
International Society for Optical Engineering 15
internet 12, 16, 42, 63, 70–72, 74, 179
Interstate Commerce Commission (ICC) 12
interviews 63, 65, 110, 111, 153
inventions 25, 72, 83, 90, 114
inventive 86, 90, 91, 93
inventors 72, 90
investment 139, 144, 179, 180
IOS 36, 152
IPC 73, 74
iron 155, 157
ISO 117
iterative 49, 56, 76, 83, 175–177

Japanese 165
Jensen 28
JPL 85

Index

Kevlar 154, 155
keywords 70, 71, 73, 75, 76, 94, 116
kinematics 118
Kydex 161

lateral 86
lathe 153
laws 11, 12, 72
layer/layering 167, 171
LDPE 161
library 15, 25, 70
licensing 11–13, 15, 72, 95
licensure 15
LinkedIn 185
literature 3, 44, 69–71, 78, 83, 84, 88, 116, 117, 138
load-level 42
local quality 93
logbook 113–116
lone-ranger 23, 84

Macs 36
magnesium 155
maintainability 17, 18, 20, 77
manufacture 64, 65, 72, 77, 78, 92, 93, 107, 110, 143, 162, 163, 173
MapleSIM 153
Markov decision tree analysis 95
Marlin 167
Mars climate orbiter 85
MATLAB 94, 95, 153
MDF 154
measurements 178
memorandum 32
meshlab 168
metals and alloys 155
microcontroller 167
Microsoft office 365 36
Microsoft office 36
Microsoft 36, 42
mild steel 155
milestones 50, 57, 111, 114, 117, 138, 144, 151, 177, 186
Milspecs 117
minerals 13, 15
mistake proofing 64, 165
MIT 90
modeling 3, 19, 22, 93, 95, 111, 118, 128, 151, 153, 167
modular 163, 164
Monte Carlo 95
morphological 90
Mostert 85
Mscloud 87
MSDS 155
multi-barrel 65
multi-function 93

multiphysics 94
musical chairs 24

nanotechnology 13
NASA 85
National Fire Protection Association (NFPA) 15
National Labor Relations Board (NLRB) 12
National Sanitary Foundation (NSF) 15
Naval Undersea Warfare Center 64
negotiation 42
nested doll 93
Nickel 155
NIST 12
non-disclosure agreement (NDA) 28
nonlinear 94, 177, 180, 181
Noryl 161
NSPE 15
Nuclear Regulatory Commission (NRC) 12
nylon 155, 161

oak 154
object-generated 92
Occupational Safety and Health Administration (OSHA) 12, 144, 155
occupational 12, 13, 155
off-the-shelf 77, 163, 164
open-ended 10, 16, 18, 21, 30, 39, 63, 67, 80, 85, 177
open-source 152, 167
operability 17, 18, 20, 77, 78
orientation 88, 131, 156, 165
Osborne 64, 86
outcome 9, 95
outside-the-box 64
outsource 173
over-constrained 176
over-design 65
overhead 6, 163, 164
overruns 32
over-the-counter 66
ownership 178

paints 155
pandemic 35, 36
paradox 10
parametric 95, 155, 177, 181
partnerships 83
parts reduction 163
patents 3, 25, 69, 72–76, 86, 92, 117, 138, 142, 167
P-card 174
PDRs 19, 114, 137, 138, 142
PEEK 160, 161
peer review 5, 18, 71
personalities 34, 84
person-hours 139, 144
PET 155
PETG 160, 161

phenolic 161
photographs/photographing 129, 130, 131, 140, 145, 185
pine 154
pins 153
pipe 158, 159
PLA 100, 101, 153, 155, 160
planner 56, 58
planning 19, 23, 39–42, 44, 110, 111, 130, 132, 138, 144, 151, 152, 175, 183
plans 3, 31, 43, 57, 65, 111, 116, 117, 129, 139, 143, 151, 152, 154
plastics 155, 157, 160, 162, 165, 167
plate 158–160, 167
pneumatics 93
POCs 109–112
poka-yoke 165, 166
polycarbonate 155, 160, 162
polymers 155–159
polypropylene 155, 162
polystyrene 155
polythene 155
POM 155
porcelain 155
porous 93
portability 77
posters 114, 118, 131, 133, 134, 135, 183–185
PowerPoint 36, 117, 120, 127–129, 131
PPS 162
practical 11, 26, 31, 39, 70, 83, 95, 109, 110, 112, 117, 180
practicing engineer 15
precision 65, 92, 164, 168
predecessors 43, 44
preliminary 3, 17, 18, 40, 94, 111, 114, 118, 137
preparation 6, 15, 24–26, 31, 41, 95, 131, 179
pressures 82, 87, 92, 147, 153
price/pricing 6, 13, 32, 80, 102, 106, 139, 143, 173
Primavera 43
print bed 167, 168
print-head 167
problem definition 10, 18, 40, 42, 63–65, 68, 69, 83–85, 97, 100, 102, 116, 117, 119, 132, 136–138, 142, 176
problem statement 10, 18, 41
problem-solver 10, 29, 30
problem-solving 3, 6, 10, 41, 63, 64, 83, 85, 86, 90, 96, 113, 180
procurement 163, 173, 174
product 3, 11, 13, 16, 25, 32, 43, 48, 65, 66, 70, 72, 74, 80, 82, 88, 90, 91, 95, 97, 99, 100, 101, 109–112, 136–139, 143, 144, 151–159, 162–165, 172, 173, 176–179, 183, 185
production 110, 118, 139, 143, 164
production-ready 172

productivity 28, 29, 92, 93
profession 8, 14, 15, 27, 35, 114, 181
professionalism 27
profits 72, 177, 179, 180
programming 22, 95, 183
progress report 23, 32, 33, 57
progression 110, 114, 129, 151
project-specific 143
ProModel 95, 151, 153
proof of concept 109, 110, 114, 139, 151
prototypes 3, 4, 7, 17, 18, 40, 77, 81, 90, 95, 109–112, 114, 137, 139, 144, 151–153, 163, 167, 184, 185, 186
prototyping 18, 107, 109–112, 164, 173
PTFE 162
public health 9, 13
publications 18, 25, 71, 78, 139, 140, 145, 146, 179
Pugh 17, 94, 96, 97, 100, 107, 128, 138, 143
purchase request 19
purchasing 19, 42, 151, 164, 173, 174
PVC 155, 162
Python 95

QFD chart 102–106, 145
qualifications 23–28
Quality Function Deployment (QFD) 17, 94, 97, 99, 100, 102–107, 110, 111, 117–119, 128, 138, 140, 143, 145
Qualtrics 65, 68
quantitative 78–80, 93, 97, 137

radar/spider charts 118
radiation 13
random input 86
ranking 97, 99
Raspberry Pi 101
rational 49
rationale 32, 144
realistic 18, 39, 48, 95
real-life 109
real-time 36
real-world 6, 10, 16, 21, 26, 42, 95, 112, 154, 175
rebar 158
recovering 93
rectangular 153, 158, 159
recycled 154
redesign 19, 20, 77, 137, 144, 149, 151, 152, 164, 172, 173, 175, 176, 179
redundant 80
reflections 116
registration 13, 15, 118, 184
regulate/regulated/regulations/regulatory 11–13, 16, 17, 19, 20, 139, 143
reliable/reliability 15, 27, 77, 82, 92, 143, 164, 178
repair 92, 144, 164, 168, 178
respect/respectful 27, 28, 44, 49, 84, 96, 173, 181

Index

responsible 3, 8, 15, 41, 69, 77, 85, 167, 174
responsiveness 29, 30
resumes 15, 22, 24, 118, 119
revolutionary 6
rework 173, 176
risk 22, 90
Ritter 85
roadmap 39
root cause 165, 175
Rubbers 156–159
rubrics 119, 176
rules of practice 11
rushing through 93
Russian 90

SAE 15, 88
safe 36, 139, 143, 154
safety 9, 11–13, 15, 17, 18, 20, 22, 65, 77, 78, 82, 118, 143, 144, 154, 155, 181
scalability 110
scaled-down 110, 111, 153
scaling 110, 144, 153
SCAMPER 86, 87
schedule/scheduling 6, 30–32, 36, 41, 42, 44, 48, 50, 108, 117, 151, 152, 154, 184–186
schematics 18, 88, 94, 113, 117, 136
screws 153
sealants 153
segmentation 93
self-check 165
self-managing 29
self-organizing 29
self-service 93
self-study 9
semester 16, 22, 44, 107, 111, 119, 137, 151, 184, 186
semester-long 44
serviceability 178
setbacks 177
seven-step strategy 73
SharePoint 36
sheets 36, 118, 136, 155, 158
short-living 93
showcase 20, 114, 183–186
signature 35, 114
silent brainstorming 84, 86
silicone 155
simulations 17, 18, 24, 93, 95, 118, 151, 153
SIMULIA 94
sinterin 167
sketches/sketching 3, 90, 114, 116, 117, 138
skills 6, 8, 9, 14, 19, 21–24, 27, 29, 30, 42, 85, 113, 163, 165, 172
Skype 33, 36
slack 53–55
slicers 167, 168

slicing 167, 168
smaller-scale 153, 154
small-scale 109
SME 15
SNAME 15
snapshot 94, 117
social 9, 35, 118, 185
societal 9, 144
Society of Automotive Engineers 15
The Society of Naval Architects and Marine Engineers 15
SolidWorks 173
specifications 3, 4, 7, 11, 16–19, 29, 31, 40, 44, 54, 65, 66, 68, 78–85, 87, 90, 91, 96, 97, 100, 101, 107, 110, 111, 116–118, 128, 131, 132, 136–139, 143, 144, 163, 175–177, 181
speed 92, 100, 101, 106, 178
spheres 93
spider charts 118
SPIE 15
sponsorship 153
square 153, 158–160
stability 92
stainless 155, 158, 159
stakeholders 27, 28, 33, 39, 48, 65, 66, 69, 112, 113, 115, 136, 137, 142, 151, 152, 153, 177, 183, 184
standards 9, 11–17, 20, 82, 117, 139, 143, 144, 178
statute 72
steel 155, 157–159, 163
step-by-step 18
stereolithography 167
stereotyping 84
stiffness 155, 156
stitch-bound 114
STL 167–170
stochastic 95
stormer 64
storming 28, 29
Stratasys 167
stress-strain 118
structural 94, 139, 143, 156, 164
sub-assemblies 173
sub-parts 88, 90
sub-sub-systems 88
subsystem 88, 143
subtracted 87
subtractive manufacturing 167
Suh 90
supply-chain 173
surveying/surveyed 66, 67
SurveyMonkey 65, 68
sustainable/sustainability 14, 17, 20, 77, 82, 118, 144, 154, 155
symposium 183, 184, 186

systematic 3
system-level 144

tactful 27
talent 29
TBD 18
tension 157
Teoriya 90
thermal 94, 139, 143, 155, 160, 162, 181
thermodynamics 22
three-ring binder 116
tips 49, 119, 122
TIS 136
Titanium 155
Title VIII 11
TMS 15
to-do list 32
tolerances/tolerancing 65, 139, 164
toolboxes 91
top-down 29
top-level 94
tough 101, 157, 158
TPU 153
trademark 72, 73
trade-off 118, 129, 138, 143
transcript 27
translation 80, 81
transparency 84
trial-and-error 180
triangles 168
TRIZ-journal 17, 86, 90–93
tube/tubing 153, 158–160, 161
Tuckman 28
tunnel vision 83, 84
Turcite 162

UHMW 162
Ultem 162
unconstrained 177
underestimate/underestimating 31, 42
uniform 93
universal serial bus 165
University of Strathclyde 96
Unix 152
unknowns 28, 109, 110, 112
unreasonable 42
unresolved 63, 110, 112, 113, 129, 175
USB 165, 168
USB-A 165, 166
USB-C 165, 166

USB-micro 166
USPTO 72–76

vagueness 63
validation 95, 128
vector 167, 181
verification/verified/verifying 95, 111, 128, 143, 163
versatility 92
vertices 167
vibration/vibrational 93, 139, 143
video-conference 33, 36, 37, 66
videography 130
videos 119, 153, 185
video-supported 131
video-teleconferencing 36
visual basic 95
visualization/visualize 88, 109, 112

wants 97, 109
wasting 65
weaknesses 9, 34, 178, 180
WebEx 33, 36
weighted 97
weights 97, 102
weld 160
welder 153
welding 42, 152, 163
whats 132
whys 132
Windows 36, 152
winning 6
wire 158
wisdom 49
wood 154–159
workarounds 39, 49, 84
workbook 114
work-in-progress 116
workload 56
workweek 56

X's 77, 78

Young's modulus 155–157

zinc 155
Zirconia 155
Zoom 33, 36
Zotero 71